Lecture Notes in Mathematics

Volume 2307

T0235641

This series reports on new developments in all areas of mathematics and their applications - quickly, informally and at a high level. Mathematical texts analysing new developments in modelling and numerical simulation are welcome. The type of material considered for publication includes:

1. Research monographs
2. Lectures on a new field or presentations of a new angle in a classical field
3. Summer schools and intensive courses on topics of current research.

Texts which are out of print but still in demand may also be considered if they fall within these categories. The timeliness of a manuscript is sometimes more important than its form, which may be preliminary or tentative.

Titles from this series are indexed by Scopus, Web of Science, Mathematical Reviews, and zbMATH.

Pierre Antoine Grillet

The Cohomology
of Commutative Semigroups

An Overview

 Springer

Pierre Antoine Grillet
Palm Bay, FL, USA

ISSN 0075-8434   ISSN 1617-9692   (electronic)
Lecture Notes in Mathematics
ISBN 978-3-031-08211-5   ISBN 978-3-031-08212-2   (eBook)
https://doi.org/10.1007/978-3-031-08212-2

Mathematics Subject Classification: 20M50, 20M14, 18G35

This Springer imprint is published by the registered company Springer Nature Switzerland AG
The registered company address is: Gewerbestrasse 11, 6330 Cham, Switzerland

*To Denise whose love, support, and understanding
made much of this monograph possible*

Preface

The cohomology of semigroups began with the extension to monoids (MacLane [41]) of the cohomology of groups and with the Leech cohomology [40]. The latter can rightly claim to be the 'right' cohomology for monoids in general [54].

We now have general ways to assign a cohomology to most algebraic structures, including commutative semigroups, thanks to Beck [6] (who used comonads), Quillen [45] (who used homotopical algebra), and André-Quillen [2, 3, 46] (who used models). All three methods yield essentially the same cohomology [45, 46]. This monograph uses Beck's, which is well suited to commutative semigroups.

An unusual feature makes the cohomology of commutative semigroups exceptionally interesting: it classifies all commutative semigroups. Indeed, the extensions it classifies are precisely those that assemble any commutative semigroup from abelian groups and a groupfree semigroup. A similar situation would exist for groups, if every group was an extension of an abelian group by a group in which every abelian normal subgroup is trivial.

Commutative semigroup cohomology has another, less felicitous feature: unwieldy cochains. Only in low dimensions do better cochains exist; how to define them in all dimensions remains an open problem.

The author's interest in the cohomology of commutative semigroups began with [18]. Some 25 years of research, including recent developments, have since matured the theory and enriched its scope and depth, making it worthy of its own monograph.

Contents are as follows. Chapter 1 outlines the construction of commutative semigroups from abelian groups and groupfree semigroups, which strongly suggests a cohomology similar to the cohomology of groups but with 'symmetry' conditions caused by commutativity.

Chapter 2 contains the basics of Beck cohomology and its adaptation to commutative semigroups.

Chapter 3 defines the author's symmetric cohomology, which is based on Chap. 1. Due to difficulties with symmetry conditions, this cohomology is only defined in dimensions $n \leq 4$, at which it coincides with Beck cohomology.

Chapter 4 outlines the cohomology developed by Calvo-Cervera and Cegarra [10] using simplicial methods, which coincides with the other two in dimensions $n \leq 2$ but yields larger groups in higher dimensions.

This monograph does not include the cohomology of Kurdiani and Pirashvili [38], which is based on derived functors; the author was unable to see how it relates to the other three. Another possible definition, as the André-Quillen cohomology [2, 3, 46] of the semigroup ring, cannot classify all commutative semigroups, for its coefficients are not sufficiently general; this is explained in Sect. 9.5.

Chapter 5 interprets H^3 as classifying extensions of sorts, following Calvo-Cervera et al. [10, 11].

Chapter 6 outlines a technique (the Overpath Method) that calculates H^2 more efficiently from presentations by generators and relations.

Chapter 7 defines groups of symmetric chains and constructs symmetric chain complexes whose cohomology is the symmetric cohomology.

Chapter 8 studies how symmetry conditions are passed on to coboundaries, and obtains likely candidates for symmetry conditions in dimensions 5 and 6.

Finally, Chap. 9 supplements the text with appendices on various topics that are mentioned without much detail in previous chapters.

The author wishes to gratefully acknowledge the help and support of his editor, Ute McCrory, and of the production team at Springer, including technical support for this Devil's invention, LaTeX.

Palm Bay, FL, USA   Pierre Antoine Grillet

Contents

List of Symbols

Lower case Roman: elements; mappings.

a_i^*	$a_1 \cdots a_{i-1}\, a_{i+1} \cdots a_n$ (Chaps. 1, 2)
d_i	Face map of simplicial set (Chap. 4, Sect. 9.3)
d_i'	Face map of V (Chap. 4, Sect. 9.3)
d_i''	Face map of W (Chap. 4)
f	A component of $\mathfrak{F}$ (Chap. 5); a symmetric mapping (Chap. 7)
$\widehat{f}$	The symmetric mapping induced by f (Chap. 7)
m	An element of M (Chap. 6)
$\overrightarrow{m}$	A defining vector (Chap. 6)
s_i	Degeneracy map of simplicial set (Chap. 4, Sect. 9.3)
s_i'	Degeneracy map of V (Chap. 4, Sect. 9.3)
s_i''	Degeneracy map of W (Chap. 4)
$\widehat{t}$	Element of Schützenberger group induced by t (Chap. 1)

Upper case Roman: sets; groups; objects.

A_n	A group of symmetric n-chains (Chap. 7)
$A_n(s; t)$	A group of symmetric n-chains (Chap. 7)
B	A convex subset of S (Chap. 7)
B_B^n	group of Beck n-coboundaries
B_e^n	Group of extended n-coboundaries (Chap. 4)
B_{NS}^n	Group of normalized symmetric n-coboundaries
B_m	Group of minimal coboundaries (Chap. 6)
B_n	Group of symmetric n-boundaries (Chap. 7)
$C(X)$	Chain group of X (Chap. 7)
C^n	Group of n-cochains
C_B^n	Group of Beck n-cochains
C_e^n	Group of extended n-cochains (Chap. 4)
C_N^n	Group of normalized n-cochains
C_{NS}^n	Group of normalized symmetric n-cochains
C_S^n	Group of symmetric n-cochains
C_m	Group of minimal cochains (Chap. 6)

C_n	Group of symmetric n-chains (Chap. 7)
C_*	Complex of symmetric n-chains (Chap. 7)
F	A left adjoint functor (Sect. 9.2)
G	An abelian group object (Chap. 2); an abelian group; a group (Sect. 9.1)
H^n	A cohomology group
H_n	A symmetric homology group (Chap. 7)
H_B^n	A Beck cohomology group
H_C^n	A Calvo-Cegarra cohomology group
H_L^n	A Leech cohomology group
H_S^n	A symmetric cohomology group
I	Identity object (Chap. 5, Sect. 9.4)
M	A monoid; a set of minimal elements (Chap. 6)
P	A symmetry property (Chap. 8)
Q, Q', Q''	Subsets of $\mathbb{F}$ (Chap. 8)
R	A cross section of $\mathcal{C}$ (Chap. 6)
$R(p, q), R(\mathfrak{p}, \mathfrak{q})$	Relations (Chap. 6)
S	A commutative semigroup or monoid
S^1	S with an identity element adjoined if needed
S'	S with an identity element forcibly adjoined
$S^{(n)}$	$S \times \cdots \times S$ (Chap. 4, 7)
T	Functor in monad (Sect. 9.2)
T, T', T''	Subsets of $\mathbb{F}$ (Chap. 8)
U	A canonical isomorphism (Chap. 7); a right adjoint functor (Sect. 9.2)
X	An object (Chap. 5); a symmetric set (Chap. 7)
X_s	A symmetric set, where $s \in S$ (Chap. 7)
$X_a^{\uparrow}$	A symmetric set, where $a \in S$ (Chap. 7)
$X_1, \ldots, X_5$	Basis of $\mathbb{F}$ (Chap. 8)
Y	An object (Chap. 5); a basis of X (Chap. 7); the standard basis of $\mathbb{F}^{(4)}$ (Chap. 8)
$Y_1, \ldots, Y_5$	Some of $X_1, \ldots, X_5$, that copy one or more equalities
Z_n	Group of symmetric n-cycles (Chap. 7)
Z_B^n	group of Beck n-cocycles
Z_e^n	Group of extended n-cocycles (Chap. 4)
Z_{NS}^n	Group of normalized symmetric n-cocycles.
Z_S^n	Group of symmetric n-cocycles
Z_m	Group of minimal cocycles (Chap. 6)

Upright upper case Roman: certain objects.

E	An extended 3-cocycle object (Chap. 5)
E($\mathcal{G}$)	Canonical split coextension by $\mathcal{G}$ (Chaps. 1, 2)
E(s)	Commutative group coextension with factor set s (Chap. 1)
M	A simplicial monoid (Sect. 9.3)
T	A symmetric 3-cocycle object (Chap. 5)

V	Classifying simplicial set (Chap. 4, Sect. 9.3)
W	Double classifying simplicial set (Chap. 4)
X	A simplicial set (Chap. 4, Sect. 9.3); *'Blackboard' bold*: fixed or explicit objects or functors
$\mathbb{A}$	Functor to abelian group valued functors (Chap. 2)
$\mathbb{B}$	Functor to Beck extensions (Chap. 2)
$\mathbb{C}$	Functor to commutative group coextensions (Chap. 2)
$\mathbb{D}$	A subgroup of $C_4(\mathbb{F})$ (Chap. 8)
$\mathbb{E}$	Functor to extended 3-cocycle objects (Chap. 5)
$\mathbb{F}$	Free commutative semigroup functor (Chap. 2); a free commutative semigroup (Chap. 6); the free commutative semigroup on $X_1, \ldots, X_5$ (Chap. 8).
$\mathbb{G}$	A free abelian group (Chap. 6)
$\mathbb{L}$	The limit functor (Chap. 4)
$\mathbb{N}$	The natural numbers
$\mathbb{O}$	Functor to abelian group objects over S (Chap. 2)
$\mathbb{P}$	The projection of Δ_W to $\mathcal{H}(S)$ (Chap. 4); the free $\mathbb{Z}$-module of symmetry properties (Chap. 8)
$\mathbb{R}$	The real numbers.
$\mathbb{T}$	Functor in commutative semigroup monad (Chap. 2); functor to symmetric 3-cocycle objects (Chap. 5)
$\mathbb{U}$	Underlying set functor (Chap. 2)
$\mathbb{V}$	Functor in commutative semigroup comonad (Chap. 2)
$\mathbb{Z}$	The integers
$\mathbb{Z}[S]$	The semigroup ring of S (Sect. 9.5)

'Euler' script: congruences; functors.

$\mathcal{A}_n$	A symmetric chain functor (Chap. 7)
$\mathcal{A}_*$	A complex of symmetric chain functors (Chap. 7)
$\mathcal{C}$	A congruence (Chaps. 1, 6)
$\mathcal{C}_n$	A thin symmetric n-chain functor (Chap. 7)
$\mathcal{C}_*$	A complex of thin symmetric chain functors (Chap. 7)
$\mathcal{F}$	A component of $\mathfrak{F}$ (Chap. 5)
$\mathcal{F}(S/B)$	The semiconstant functor on B at G (Chap. 7)
$\mathcal{G}$	An abelian group valued functor on S
$\mathcal{G}'$	$\mathcal{G}$ extended to S' (Chap. 3)
$\mathcal{G}'_\varphi$	$\mathcal{G}' \circ \mathcal{H}(\varphi)$ (Chap. 5)
$\mathcal{H}$	The divisibility congruence on S (Chap. 1)
$\mathcal{S}$	An exact sequence (Chaps. 2, 3, 4)

'Calligraphic' script: categories.

$\mathcal{A}$	Category of all abelian group valued functors (Chap. 7)
$\mathcal{B}$	Category of reduced braided monoidal abelian groupoids (Chap. 5)
$\mathcal{B}(S, \mathcal{G})$	A subcategory of $\mathcal{B}$ (Chap. 5)

$\mathcal{C}$	A category, perhaps the cat. of commutative semigroups (Chap. 2); a small category (Chap. 4)
$\mathcal{D}(S)$	One of the Leech categories of S
$\mathcal{E}$	Category of extended 3-cocycle objects (Chap. 5)
$\mathcal{E}(S, \mathcal{G})$	A subcategory of $\mathcal{E}$ (Chap. 5)
$\mathcal{G}$	A groupoid (Chap. 5)
$\mathcal{H}(S)$	One of the Leech categories of S
$\mathcal{H}(\varphi)$	Functor induced by φ (Chap. 5)
$\mathcal{L}(S)$	One of the Leech categories of S (Sect. 9.1).
$\mathcal{M}$	An abelian groupoid (Chap. 5)
$\mathcal{R}$	Category of reduced comm. monoidal abelian groupoids (Chap. 5)
$\mathcal{R}(S)$	One of the Leech categories of S (Sect. 9.1)
$\mathcal{R}(S, \mathcal{G})$	Category of reduced comm. monoidal abelian groupoids; with base $(S, \mathcal{G})$ (Chap. 5)
$\mathcal{T}$	Category of symmetric 3-cocycle objects (Chap. 5); category of thin abelian group valued functors (Chap. 7)
$\mathcal{T}(S, \mathcal{G})$	A subcategory of $\mathcal{T}$ (Chap. 5)
$\mathcal{X}$	A category, perhaps the category of sets (Chap. 2)

Barred letters:

$\overline{C}$	An object over S (Chap. 2)
$\overline{\mathcal{C}}$	Category of objects over S (Chap. 2)
$\overline{\mathbb{F}}$	$\mathbb{F}$ lifted to $\overline{\mathcal{C}}$ (Chap. 2)
$\overline{G}$	An abelian group object over S (Chap. 2)
$\overline{p}$	$\varepsilon_S \circ \mathbb{F}p$ (Chap. 2)
$\overline{P}_a$	A probe (Chap. 2)
$\overline{S}$	$(S, 1_S)$ (Chap. 2)
$\overline{\mathbb{T}}$	$\mathbb{T}$ lifted to $\overline{\mathcal{C}}$ (Chap. 2)
$\overline{T}_n$	Iterated free commutative semigroup (Chap. 2)
$\overline{\mathbb{U}}$	$\mathbb{U}$ lifted to $\overline{\mathcal{C}}$ (Chap. 2)
$\overline{\mathbb{V}}$	$\mathbb{V}$ lifted to $\overline{\mathcal{C}}$ (Chap. 2)
$\overline{\mathcal{X}}$	Category of objects over $\mathbb{U}S$ (Chap. 2)
$\overline{\varepsilon}$	ε lifted to $\overline{\mathcal{C}}$ (Chap. 2)
$\overline{\eta}$	η lifted to $\overline{\mathcal{C}}$ (Chap. 2)
$\overline{\pi}$	$\varepsilon_S \circ \mathbb{V}\pi$ (Chap. 2)

Lower case Greek: homomorphisms and natural transformations (mostly).

γ	A homomorphism in an abelian group valued functor
γ_K^H	A Schützenberger homomorphism (Chap. 1, Sect. 9.1)
δ	Coboundary homomorphism
δ_i	A face morphism of Δ (Chap. 4, Sect. 9.3)
δ^i	A face homomorphism of Calvo-Cegarra cochains (Chap. 4)
ε	Natural transformation in adjunction or comonad (Chap. 2, Sect. 9.2)

$\varepsilon^{n,i}$	Natural transformation derived from ε (Chap. 2)
$\varepsilon(x)$	Least idempotent e such that $ex = x$ (Chap. 1)
ζ	A component of $\mathfrak{F}$ (Chap. 5)
η	Natural transformation in adjunction (Chap. 2, Sect. 9.2)
θ	An isomorphism
ι	Canonical mapping $X \longrightarrow C(X)$ or $S^{(n)} \longrightarrow C_n(S)$ (Chap. 7)
μ	Natural transformation in monad (Sect. 9.2)
ν	Natural transformation in comonad (Chap. 2, Sect. 9.2)
π	Projection to S in coextension (Chap. 1, Sect. 9.1)
σ	A permutation
σ_i	A degeneracy morphism of Δ (Chap. 4, Sect. 9.3)
σ^i	A degeneracy homomorphism of Calvo-Cegarra cochains (Chap. 4)
σ_*	Homomorphism induced by natural transformation σ (Chap. 5)
τ^φ	Natural transformation induced by τ and φ (Chap. 5)
φ^*	Homomorphism induced by φ (Chap. 5)

Upper case Greek.

Δ	The simplicial category (Chap. 4, Sect. 9.3)
Δ^a	The augmented simplicial category (Sect. 9.3)
Δ_X	Category of simplexes of X (Chap. 4)
Φ, Ψ	Morphisms of cocycle objects (Chap. 5)

Lower case German 'fraktur': structural isomorphisms; paths.

$\mathfrak{a}$	'Associativity' isomorphism (Chap. 5, Sect. 9.4)
$\mathfrak{c}$	'Commutativity' or braiding isomorphism (Chap. 5, Sect. 9.4)
$\mathfrak{d}$	The universal 5-coboundary (Chap. 8)
$\mathfrak{i}$	'Identity' isomorphism (Chap. 5, Sect. 9.4); the canonical mapping $\mathbb{F}^{(4)} \longrightarrow C_4(\mathbb{F})$ (Chap. 8)
$\mathfrak{l}$	'Left identity' isomorphism (Chap. 5, Sect. 9.4)
$\mathfrak{p}, \mathfrak{q}$	Paths in $\mathbb{F}$ (Chap. 6)
$\mathfrak{p}^t$	Translated path (Chap. 6)
$\mathfrak{r}$	'Right identity' isomorphism (Chap. 5, Sect. 9.4)

Upper case German 'fraktur': monoidal categories and morphisms.

$\mathfrak{F}$	A monoidal functor (Sect. 9.4.); a morphism of monoidal abelian groupoids (Chap. 5, Sect. 9.4)
$\mathfrak{M}$	A monoidal category (Sect. 9.4); a monoidal abelian groupoid (Chap. 5, Sect. 9.4)

Miscellaneous.

$(x_1, \ldots, x_n)$	Ordered sequence
$[x_1, \ldots, x_n]$	Unordered sequence
$\langle a_1, \ldots, a_n \rangle$	Symmetric chain (Chaps. 7, 8)
$\mathcal{C} \downarrow S$	Category of objects over S (Chap. 2).

∂	Boundary homomorphism or natural transformation (Chap. 7)
$\cdot$	Group action in coextension (Chap. 1); group action in Beck extension (Chap. 2); product in category (Chap. 5, Sect. 9.4)
$\leq$	The divisibility preorder on S; in Chap. 6, the natural order on $\mathbb{F}$; in Chap. 7, an arbitrary total order on S; in Chap. 8, a total order on $\mathbb{F}$
$\prec$	A compatible well order on $\mathbb{F}$ (Chap. 6)
$(\mathfrak{p}, \mathfrak{q})$	A branching pair (Chap. 6)
$a \xrightarrow{m} b$	An edge in $\mathbb{F}$ (Chap. 6)

Chapter 1
The Beginning

The cohomology of commutative semigroups, like the cohomology of groups, arose from fundamental considerations of structure.

1.1 The Congruence $\mathcal{H}$

1.1.1 Basics

First we recall a few basic facts about commutative semigroups (from e.g. [13] or [31]).

As a rule, we write semigroup operations as multiplications.

A *monoid* is a semigroup M that has an identity element 1. A semigroup S that does not have an identity element is readily completed to a monoid $S^1 = S \cup \{1\}$, with the multiplication induced by S. If S already has an identity element, then $S^1 = S$. If S is commutative, then so is S^1.

An *idempotent* of S is an element e such that $e^2 = e$.

A *congruence* on a semigroup S is an equivalence relation $\mathcal{C}$ such that $a \ \mathcal{C} \ b$ implies $sa \ \mathcal{C} \ sb$ and $as \ \mathcal{C} \ bs$ for all $s \in S$. The projection $\pi : S \longrightarrow S/\mathcal{C}$ to the quotient set then induces a multiplication on $S/\mathcal{C}$ such that π is a homomorphism. If S is commutative, or a monoid, then so is $S/\mathcal{C}$.

Divisibility on a commutative semigroup S is a preorder relation (reflexive and transitive):

$$a \leq b \ \text{if and only if} \ a = bs \ \text{for some} \ s \in S^1.$$

To this divisibility preorder corresponds an equivalence relation:

$$a \ \mathcal{H} \ b \ \text{if and only if} \ a \leq b \ \text{and} \ b \leq a,$$

© The Author(s), under exclusive license to Springer Nature Switzerland AG 2022
P. A. Grillet, *The Cohomology of Commutative Semigroups*, Lecture Notes in Mathematics 2307, https://doi.org/10.1007/978-3-031-08212-2_1

which is a congruence on S. The $\mathcal{H}$-classes of the idempotents of S are the maximal subgroups of S [13].

A commutative semigroup S is *groupfree* if and only if $\mathcal{H}$ is the equality on S; equivalently, all the Schützenberger groups of S (defined below) are trivial. In particular, all subgroups of S are then trivial. For every commutative semigroup S the quotient semigroup $S/\mathcal{H}$ is groupfree; it is the *groupfree semigroup* of S. In particular, the divisibility preorder on $S/\mathcal{H}$ is a partial order relation.

Example 1 Let S be the commutative monoid $\{1, a, b, c, e, 0\}$ with multiplication

1	a	b	c	e	0
a	1	c	b	0	0
b	c	0	0	0	0
c	b	0	0	0	0
e	0	0	0	e	0
0	0	0	0	0	0

There are four $\mathcal{H}$-class es: $\{1, a\}$, $\{b, c\}$, $\{e\}$, and $\{0\}$. The groupfree semigroup of S is isomorphic to

1	b	e	0
b	0	0	0
e	0	e	0
0	0	0	0

□

Let H be an $\mathcal{H}$-class of S. Every $t \in S^1$ such that $tH \subseteq H$ induces a mapping $\widehat{t_H} : H \longrightarrow H$, $x \longmapsto tx$. The mappings $\widehat{t_H}$ constitute a simply transitive group of permutations of H, the *Schützenberger group* $G(H)$ of H [13].

The Schützenberger groups of S are connected by homomorphisms. If $H \geq K$ in $S/\mathcal{H}$, then $Hs \subseteq K$ for some $s \in S^1$, and $tH \subseteq H$ implies $tK \subseteq K$. There is a well defined *Schützenberger homomorphism*

$$\gamma = \gamma_K^H : G(H) \longrightarrow G(K)$$

that sends $\widehat{t_H}$ to $\widehat{t_K}$. If $g = \widehat{t} \in G(H)$, then $g(x) = tx$ and

$$g(x)s = (tx)s = t(xs) = \gamma(g)(xs) \tag{1.1}$$

for all $x \in H$. Also, γ_H^H is the identity on $G(H)$ and $\gamma_L^K \circ \gamma_K^H = \gamma_L^H$ whenever $H \geq K \geq L$ in $S/\mathcal{H}$. Hence the Schützenberger groups and homomorphisms constitute an abelian group valued functor, called the *Schützenberger functor* of S, on the partially ordered set $S/\mathcal{H}$, when the latter is viewed as a category, in which an object is an element of $S/\mathcal{H}$ and there is one morphism $(a, b)\colon a \to b$ for each pair (a, b) such that $a \geq b$ in $S/\mathcal{H}$.

Example 1 is too small to have interesting Schützenberger groups: the Schützenberger groups of $\{1, a\}$ and $\{b, c\}$ are cyclic of order 2; the Schützenberger groups of $\{e\}$ and $\{0\}$ are trivial. If the $\mathcal{H}$-class es are renamed $1, b, e, 0$, then γ_b^1 is an isomorphism; the other homomorphisms are identities or trivial.

In view of Theorem 2.2.1, abelian group valued functors on a commutative semigroup S are best defined as functors on the *Leech category* $\mathcal{H}(S)$, whose objects are the elements of S and whose morphisms are the elements $(a, t)\colon a \longrightarrow at$ of $S \times S^1$, composed by $(at, u) \circ (a, t) = (a, tu)$, so that $(a, 1)$ is the identity on a. Thus an *abelian group valued functor* $\mathcal{G} = (G, \gamma)$ *on* S (also called an $\mathcal{H}(S)$-*module* [10]) assigns an abelian group G_a to each $a \in S$, and a homomorphism $\gamma_{a,t}\colon G_a \longrightarrow G_{at}$ to each $a \in S$ and $t \in S^1$, so that $\gamma_{a,1}$ is the identity on G_a and $\gamma_{at,u} \circ \gamma_{a,t} = \gamma_{a,tu}$.

It is convenient to denote $\gamma_{a,t}g$ by g^t; then $g^1 = g$ and $(g^t)^u = g^{tu}$ for all $t, u \in S^1$, whenever $g \in G_a$ for some $a \in S$.

An abelian group valued functor $\mathcal{G} = (G, \gamma)$ on $\mathcal{H}(S)$ is *thin* if and only if $\gamma_{a,t} = \gamma_{a,u}$ whenever $at = au$. If $\mathcal{G} = (G, \gamma)$ is thin, then γ yields only one homomorphism $\gamma_b^a\colon G_a \longrightarrow G_b$ when $a \geq b$, and then γ_a^a is the identity on G_a and $\gamma_c^b \circ \gamma_b^a = \gamma_c^a$ whenever $a \geq b \geq c$. Thus $\mathcal{G}$ can be viewed as a functor on the preordered set $(S, \leq)$. Conversely, an abelian group valued functor $\mathcal{G} = (G, \gamma)$ on the preordered set $(S, \leq)$ can be viewed as a thin abelian group valued functor on $\mathcal{H}(S)$ in which $\gamma_{a,t} = \gamma_{at}^a$. For example, all Schützenberger functors are thin (as functors on $\mathcal{H}(S)$).

1.1.2 Commutative Group Coextensions

A *commutative group coextension* E or $(E, \pi, \cdot)$ of a commutative semigroup S by an abelian group valued functor $\mathcal{G} = (G, \gamma)$ on S consists of a commutative semigroup E, a surjective homomorphism $\pi\colon E \longrightarrow S$ which partitions E into sets

$$E_a = \{x \in E \mid \pi x = a\},$$

and, for each $a \in S$, a simply transitive action $\cdot$ of G_a on E_a, such that

$$(g \cdot x)\, y = g^b \cdot xy \tag{1.2}$$

for all $a, b \in S$, $x \in E_a$, and $y \in E_b$.

The author uses the ugly word *coextension* because he could not bring himself to view the functor $\mathcal{G}$ as a subobject of the semigroup E. Other authors call E an *extension* of $\mathcal{G}$ by S. See also Part 2.1.6.

Commutative group coextensions include abelian extensions of abelian groups. Indeed let B be a subgroup of an abelian group A. Let $\mathcal{G}$ be the *constant* functor on

$Q = A/B$ that assigns B to every $q \in Q$ and the identity on B to every $q, r \in Q$. Then A is a commutative group coextension of Q by $\mathcal{G}$, in which B acts on its cosets by multiplication.

Comparing (1.1) and (1.2) yields:

Proposition 1.1.1 *Every commutative semigroup is a commutative group coextension of its groupfree semigroup by its Schützenberger functor.*

Two commutative group coextensions $(E, \pi, \cdot)$ and $(F, \rho, \cdot)$ of S by $\mathcal{G}$ are *equivalent* if and only if there exists an isomorphism $\theta : E \longrightarrow F$ that preserves projection to S ($\rho \circ \theta = \pi$) and preserves the action of $\mathcal{G}$ ($\theta (g \cdot x) = g \cdot \theta(x)$, for all $a \in S$, $x \in E_a$, and $g \in G_a$).

For example, two abelian group extensions of A by Q are equivalent as group extensions if and only if they are equivalent as group coextensions.

1.2   Construction

1.2.1   Schreier's Method

The central fact of this chapter is that S can be reconstructed from its underlying groupfree semigroup and Schützenberger functor.

Group coextensions are very similar to group extensions and are constructed in much the same way, as in [52]. In fact, both constructions are particular cases of the more general constructions in [18] and [39] (see Sect. 9.1).

Let $(E, \pi, \cdot)$ be a commutative group coextension of S by $\mathcal{G}$. For each $a \in S$ choose $p_a \in E_a$. Since G_a acts simply and transitively on E_a, every element x of E can be written in the form

$$x = g \cdot p_a$$

for some unique $a \in S$ and $g \in G_a$. In particular,

$$p_a \, p_b = s_{a,b} \cdot p_{ab} \text{ for some unique } s_{a,b} \in G_{ab}. \tag{1.3}$$

The family $s = (s_{a,b})_{a,b \in S}$ is the *factor set* of E relative to the cross-section p. This terminology goes back to Schreier [52].

The multiplication on E is completely determined by S, $\mathcal{G}$, and s: indeed, Eqs. (1.2), (1.3) and commutativity yield

$$(g \cdot p_a)(h \cdot p_b) = (g^b \, s_{a,b} \, h^a) \cdot p_{ab}$$

for all $a, b \in S$, $g \in G_a$, and $h \in G_b$.

The factor set s has two properties that reflect associativity and commutativity in E:

$$s_{b,c}^a \, s_{a,bc} = s_{ab,c} \, s_{a,b}^c \tag{1.4}$$

$$s_{b,a} = s_{a,b} \tag{1.5}$$

for all $a, b, c \in S$. Property (1.5), which looks so innocent, is in fact the cause of major unanswered questions in this monograph.

Conversely, (1.4) and (1.5) ensure associativity and commutativity. Indeed, given a commutative semigroup S, an abelian group valued functor $\mathcal{G} = (G, \gamma)$ on S, and $s_{a,b} \in G_{ab}$ for every $a, b \in S$, let

$$E(s) = \{(g, a) \mid g \in G_a, a \in S\}$$

with multiplication

$$(g, a)(h, b) = (g^b \, s_{a,b} \, h^a, \, ab), \tag{1.6}$$

projection $(g, a) \longmapsto a$ to S, and action $g \cdot (h, a) = (gh, a)$ of each G_a on E_a.

Theorem 1.2.1 ([18]) *If S is a commutative semigroup, $\mathcal{G} = (G, \gamma)$ is an abelian group valued functor on S, $s_{a,b} \in G_{ab}$ and $s = (s_{(a,b)})_{a,b \in S}$ has properties (1.4) and (1.5) for all $a, b, c \in S$, then $\mathrm{E}(s)$ is a commutative group coextension of S by $\mathcal{G}$. Conversely, every commutative group coextension of S by $\mathcal{G}$ with factor set s is equivalent to $\mathrm{E}(s)$.*

Moreover, two commutative group coextensions of S by $\mathcal{G} = (G, \gamma)$ with factor sets s and t are equivalent if and only if there exists $u = (u_a)_{a \in S}$ such that $u_a \in G_a$ for all $a \in S$ and

$$t_{a,b} = s_{a,b} \, u_b^a \, u_{ab}^{-1} \, u_a^b \tag{1.7}$$

for all $a, b \in S$.

Using Theorem 1.2.1, one can compute all equivalence classes of commutative group coextensions of a given commutative semigroup S by a given abelian group valued functor $\mathcal{G}$ on S. The amount of work involved, however, increases very quickly with the number of elements of S and of the groups G_a. For the groupfree semigroup and Schützenberger functor of Example 1 this will be carried out in Chap. 3 and, using more efficient methods, in Chap. 6.

1.2.2 Split Coextensions

Some group coextensions are easier to construct than others.

A commutative group coextension *splits* if and only if it can be arranged that $s_{a,b} = 0$ for all $a, b \in S$; equivalently, there is a homomorphism $a \longmapsto p_a$ that splits the projection to S.

The commutative semigroup S in Example 1 is a split coextension of its groupfree semigroup: the multiplication tables show an obvious homomorphism $S/\mathcal{H} \longrightarrow S$ that splits the projection $S \longrightarrow S/\mathcal{H}$.

With $s = 0$, Theorem 1.2 yields for each abelian group valued functor $\mathcal{G}$ on S a split coextension $E(\mathcal{G}) = E(0) = (E, \rho, \cdot)$ of S by $\mathcal{G}$, the *canonical split coextension* of S by $\mathcal{G}$, in which

$$E = \{(g, a) \mid g \in G_a, a \in S\}$$

with multiplication

$$(g, a)(h, b) = (g^b\, h^a,\, ab),$$

projection $\rho : (g, a) \longmapsto a$ to S, and action $g \cdot (h, a) = gh \cdot a$ of each G_a on E_a. By Theorem 1.2, a group coextension of S by $\mathcal{G}$ splits if and only if it is equivalent to $E(\mathcal{G})$.

For future reference we look at longer products in $E(\mathcal{G})$. First,

$$(g, a)(h, b)(k, c) = (g^b\, h^a,\, ab)(k, c) = (g^{bc}\, h^{ac}\, k^{ab},\, abc).$$

In general

Lemma 1.2.1 *Let $\mathcal{G}$ be an abelian group valued functor on S and $E = E(\mathcal{G})$ be the canonical split coextension. If $n > 0$ and $(g_i, a_i) \in E$ for all i, then*

$$(g_1, a_1)(g_2, a_2) \ldots (g_n, a_n) = \Big(\prod_{1 \le i \le n} g_i^{a_i^*},\, a \Big), \tag{1.8}$$

where $a = a_1 a_2 \cdots a_n$ and

$$a_i^* = \begin{cases} a_1 \cdots a_{i-1} a_{i+1} \cdots a_n & \text{if } 1 < i < n, \\ a_2 \cdots a_n & \text{if } i = 1 \text{ and } n > 1, \\ 1 \in S^1 & \text{if } i = 1 \text{ and } n = 1, \\ a_1 \cdots a_{n-1} & \text{if } i = n \text{ and } n > 1. \end{cases}$$

A common notation for a_i^* is $a_1 \cdots \widehat{a_i} \cdots a_n$.

1.2.3 Enter Cohomology

So far we have written the abelian groups G_a multiplicatively. When these groups are written additively, as they will be in all subsequent chapters, (1.6) becomes

$$(g, a)(h, b) = (g^b + s_{a,b} + h^a, \ ab);$$

(1.4) and (1.7) become

$$s_{b,c}^a - s_{ab,c} + s_{a,bc} - s_{a,b}^c = 0 \tag{1.9}$$

$$t_{a,b} - s_{a,b} = u_b^a - u_{ab} + u_a^b; \tag{1.10}$$

and it becomes clear that factor sets are, in a suitable sense, 2-cocycles; that two factor sets are equivalent if and only if they differ by a 2-coboundary; that there is a cohomology that *classifies* commutative group coextensions in the sense that there is a one-to-one correspondence between the elements of the cohomology group H^2 and the equivalence classes of commutative group coextensions; and that we have a generous hint at its definition, which will be pursued in Chap. 3.

1.2.4 Finite Semigroups

Classifying group coextensions, however, is not quite the same as classifying the reconstruction of commutative semigroups from groupfree semigroups and abelian groups. This is because a group coextension E of S by $\mathcal{G}$ does not necessarily have S as its groupfree semigroup, or $\mathcal{G}$ as its Schützenberger functor, even if S is groupfree. This difficulty vanishes when the semigroups are finite.

The Schützenberger functors of finite commutative semigroups have additional properties. When S is finite there is for each $x \in S$ a least idempotent $e \in S^1$ such that $ex = x$: namely, the product $\varepsilon(x)$ of all such idempotents. This also applies to the finite semigroup $S/\mathcal{H}$.

Proposition 1.2.1 ([18]) *Let S be a finite commutative semigroup and let $H \in S/\mathcal{H}$ be an $\mathcal{H}$-class of S. If $S/\mathcal{H}$ does not have an identity element and $\varepsilon(H) = 1 \notin S/\mathcal{H}$, then $G(H) = \{1\}$. If $\varepsilon(H) \in S/\mathcal{H}$, then the Schützenberger homomorphism $G(\varepsilon(H)) \longrightarrow G(H)$ is surjective.*

In particular, if S is finite, then $\mathcal{H}$ is the equality on S if and only if all subgroups of S are trivial (hence the term groupfree).

An abelian group valued functor $\mathcal{G} = (G, \gamma)$ on a finite commutative semigroup S is *surjecting* if and only if it has the properties in Prop. 1.2.1: for every $a \in S$, if $\varepsilon(a) \in S$, then $\gamma_{\varepsilon(a), a} \colon G_{\varepsilon(a)} \longrightarrow G_a$ is surjective; otherwise (if S does not have an identity element and $\varepsilon(a) = 1 \notin S$) $G_a = 0$.

Proposition 1.2.2 ([18]) *The Schützenberger functor of a finite commutative semigroup is thin and surjecting. Conversely, if E is a commutative group coextension of a finite, groupfree, commutative semigroup S by a thin, surjecting abelian group valued functor $\mathcal{G}$ on S, then $E/\mathcal{H} \cong S$ and, up to isomorphisms, $\mathcal{G}$ is the Schützenberger functor of E.*

Hence a cohomology of S that classifies commutative group coextensions of S by $\mathcal{G}$ also classifies finite commutative semigroups, when S is finite groupfree and $\mathcal{G}$ is thin and surjecting.

The reach of this classification extends beyond finite semigroups. A finitely generated commutative semigroup S is *complete* if and only if every archimedean component of S contains an idempotent [19] (a more general definition is given in [31]); then $S/\mathcal{H}$ is finite as well as groupfree. Prop. 1.5 extends to finitely generated complete semigroups [31, Theorem V.3.4], so that commutative semigroup cohomology also classifies these semigroups. Commutative semigroup cohomology even reaches all finitely generated commutative semigroups, to some extent, since they can be embedded into complete semigroups [19]. And, of course, the classification of commutative group coextensions applies to all commutative semigroups, but is not in general as interesting a structure tool since $\mathcal{H}$ tends to be small unless S is finite or complete.

Chapter 2
Beck Cohomology

Beck [6] devised a cohomology with good properties for every object of a category $\mathcal{C}$ that is blessed with an adjunction into $\mathcal{C}$ (see also Barr and Beck [5]). This cohomology was called *triple cohomology* in [31] and various papers, after the title of [6]. It now seems more appropriate to the author to name it after Beck.

Particular cases of Beck cohomology include every cohomology that was of algebraic importance at the time [6, 15], as well as the Leech cohomology of monoids in general [39, 40, 54] (defined in Sect. 9.1). In the case of commutative semigroups, it also classifies commutative group coextensions. This gives Beck cohomology a strong claim to be the 'right' cohomology for commutative semigroups in general, notwithstanding severely overweight cochains.

2.1 General Beck Cohomology

The results in this section are due to Beck [6] (see also [5]).

2.1.1 Simple Cohomology

Beck cohomology arises from the following construction. Let (V, ε, ν) be a comonad on a category $\mathcal{C}$. For each $0 \leq i \leq n$ the natural transformation $\varepsilon \colon V \longrightarrow 1_{\mathcal{C}}$ induces natural transformations

$$\varepsilon^{n,i} \;=\; V^{n-i}\,\varepsilon\,V^i \colon V^{n+1} \longrightarrow V^n \,.$$

© The Author(s), under exclusive license to Springer Nature Switzerland AG 2022
P. A. Grillet, *The Cohomology of Commutative Semigroups*, Lecture Notes
in Mathematics 2307, https://doi.org/10.1007/978-3-031-08212-2_2

It is readily verified that

$$\varepsilon^{n,j} \circ \varepsilon^{n+1,i} = \varepsilon^{n,i} \circ \varepsilon^{n+1,j+1} \tag{2.1}$$

for all $0 \leq i \leq j \leq n$.

Given a contravariant abelian group valued functor A on $\mathcal{C}$, a chain complex of abelian groups can now be assigned to every object C of $\mathcal{C}$:

$$0 \longrightarrow AVC \longrightarrow \ldots \longrightarrow AV^n C \xrightarrow{\delta_n} AV^{n+1}C \longrightarrow \ldots \tag{2.2}$$

where

$$\delta_n = \sum_{0 \leq i \leq n} (-1)^i A\varepsilon_C^{n,i}. \tag{2.3}$$

It follows from (2.1) that

$$\delta_{n+1} \circ \delta_n = 0. \tag{2.4}$$

The cohomology groups of C with coefficients in A are the homology groups of the complex (2.2).

Underlying this construction is an augmented cosimplicial abelian group (see Part 9.3.3 or [42]); (2.1) is part of its simplicial identities.

2.1.2   Abelian Group Objects

An *abelian group object* in a category $\mathcal{C}$ is an object G such that $\text{Hom}_{\mathcal{C}}(-, G)$ is a contravariant abelian group valued functor on $\mathcal{C}$: in more detail, such that every $\text{Hom}_{\mathcal{C}}(C, G)$ is a set, together with an addition on each $\text{Hom}_{\mathcal{C}}(C, G)$ that makes $\text{Hom}_{\mathcal{C}}(C, G)$ an abelian group and distributes composition:

$$(u + v) \circ \varphi = (u \circ \varphi) + (v \circ \varphi)$$

for every $\varphi: D \longrightarrow C$ and $u, v: C \longrightarrow G$. (This definition is from [6]. It is not mentioned in [42] but, for Beck cohomology, it's just what the doctor ordered.)

A *morphism* $\psi: G \longrightarrow H$ of abelian group objects is a morphism of $\mathcal{C}$ such that $\text{Hom}_{\mathcal{C}}(C, \psi)$ is a homomorphism for every object C:

$$\psi \circ (u + v) = (\psi \circ u) + (\psi \circ v)$$

for every $u, v: C \longrightarrow G$.

For example, (ordinary) abelian groups are abelian group objects in the category of sets; a homomorphism of abelian groups is a morphism of abelian group objects.

With $A = \mathrm{Hom}_{\mathcal{C}}(-, G)$, where G is an abelian group object of $\mathcal{C}$, (2.2) becomes

$$0 \longrightarrow \mathrm{Hom}_{\mathcal{C}}(VC, G) \cdots \mathrm{Hom}_{\mathcal{C}}(V^n C, G) \xrightarrow{\delta_n} \mathrm{Hom}_{\mathcal{C}}(V^{n+1} C, G) \cdots \qquad (2.5)$$

The cohomology groups of C with coefficients in G are the homology groups of the complex (2.5):

$$H^n(C, G) = \mathrm{Ker}\,\delta_n \,/\, \mathrm{Im}\,\delta_{n-1} \text{ if } n \geq 2, \text{ with } H^1(C, G) = \mathrm{Ker}\,\delta_1\,.$$

It would be more logical to call these groups H^{n-1} and H^0, as do [6] and [5], but this would eliminate the traditional association of H^2 with extensions.

2.1.3 Objects Over S

The Beck cohomology of an object S of $\mathcal{C}$ is obtained when the above is applied to the 'comma' category of objects over S in $\mathcal{C}$, denoted here by $\overline{\mathcal{C}}$ or by the more common notation $\mathcal{C} \downarrow S$.

As in [42], an *object over* S in $\mathcal{C}$ (an object of $\overline{\mathcal{C}}$) is a pair (C, π) of an object C of $\mathcal{C}$ and a morphism $\pi : C \longrightarrow S$. A morphism of $\overline{\mathcal{C}}$ from (C, π) to (D, ρ) is a morphism $\varphi : C \longrightarrow D$ of $\mathcal{C}$ such that $\rho \circ \varphi = \pi$.

Any functor $V : \mathcal{C} \longrightarrow \mathcal{C}$ and natural transformation $\varepsilon : V \longrightarrow 1_{\mathcal{C}}$ lift to a functor $\overline{V} : \overline{\mathcal{C}} \longrightarrow \overline{\mathcal{C}}$ and natural transformation $\overline{\varepsilon} : \overline{V} \longrightarrow 1_{\overline{\mathcal{C}}}$, where $\overline{\varepsilon}_{(C, \pi)} = \varepsilon_C$,

$$\overline{V}(C, \pi) = (VC, \overline{\pi}), \text{ where } \overline{\pi} = \varepsilon_S \circ V\pi\,, \qquad (2.6)$$

and $\overline{V}\varphi = V\varphi$, for every object (C, π) and morphism φ of $\overline{\mathcal{C}}$.

If V and ε arise from an adjunction, then so do $\overline{V}$ and $\overline{\varepsilon}$. Indeed, an adjunction $(F, U, \eta, \varepsilon)$ from another category $\mathcal{X}$ to $\mathcal{C}$ lifts to an adjunction $(\overline{F}, \overline{U}, \overline{\eta}, \overline{\varepsilon})$ from $\overline{\mathcal{X}}$ to $\overline{\mathcal{C}}$, where $\overline{\mathcal{X}} = \mathcal{X} \downarrow US$ is the category of objects over US in $\mathcal{X}$,

$$\overline{F}(X, p) = (FX, \overline{p}), \text{ where } \overline{p} = \varepsilon_S \circ Fp$$

is the morphism $\overline{p} : FX \longrightarrow S$ such that $U\overline{p} \circ \eta_X = p$,

$$\overline{U}(C, \pi) = (UC, U\pi),$$

$$\overline{\eta}_{(X, p)} = \eta_X : (X, p) \longrightarrow (FX, \overline{p}), \text{ and}$$

$$\overline{\varepsilon}_{(C, \pi)} = \varepsilon_C : \overline{F}\,\overline{U}(C, \pi) \longrightarrow (C, \pi)\,.$$

If $V = FU$, then the above yield

$$\overline{FU}(C, \pi) = (VC, \overline{\pi}), \quad \text{where } \overline{\pi} = \varepsilon_S \circ V\pi,$$

so that $\overline{F}\,\overline{U} = \overline{V}$.

2.1.4 Beck Cohomology

Given V, ε, S, and an abelian group object $\overline{G} = (G, \rho)$ of $\overline{C}$, we now have for each object $\overline{C} = (C, \pi)$ of $\overline{C}$ a chain complex of abelian groups

$$0 \longrightarrow \operatorname{Hom}_{\overline{C}}(\overline{V}\,\overline{C}, \overline{G}) \cdots \operatorname{Hom}_{\overline{C}}(\overline{V}^n\overline{C}, \overline{G}) \xrightarrow{\delta_n} \operatorname{Hom}_{\overline{C}}(\overline{V}^{n+1}\overline{C}, \overline{G}) \cdots \quad (2.7)$$

Since

$$\overline{\varepsilon}_{\overline{C}}^{n,i} = \overline{V}^{n-i}\,\overline{\varepsilon}\,\overline{V}^i = \varepsilon_C^{n,i} : V^{n+1}C \longrightarrow V^n C,$$

it follows that

$$\delta_n = \sum_{0 \le i \le n} (-1)^i \operatorname{Hom}_{\overline{C}}(\overline{\varepsilon}_{\overline{C}}^{n,i}, \overline{G}) \quad (2.8)$$

takes $u : \overline{V}^n\overline{C} \longrightarrow \overline{G}$ to

$$\delta_n u = \sum_{0 \le i \le n} (-1)^i (u \circ \varepsilon_C^{n,i}). \quad (2.9)$$

The homology groups of (2.7) are the *Beck cohomology groups of* $\overline{C}$ with coefficients in $\overline{G}$:

$$H_B^n(\overline{C}, \overline{G}) = \operatorname{Ker} \delta_n / \operatorname{Im} \delta_{n-1} \text{ if } n \ge 2, \text{ with } H_B^1(\overline{C}, \overline{G}) = \operatorname{Ker} \delta_1.$$

The *Beck cohomology groups of* S itself are those of $\overline{S} = (S, 1_S)$:

$$H_B^n(S, \overline{G}) = H_B^n(\overline{S}, \overline{G}).$$

2.1.5 Main Properties

The main properties of Beck cohomology require that V and ε arise from an adjunction $(F, U, \eta, \varepsilon)$ from another category $\mathcal{X}$ to $\mathcal{C}$, which lifts as above to an adjunction $(\overline{F}, \overline{U}, \overline{\eta}, \overline{\varepsilon})$ from $\overline{\mathcal{X}}$ to $\overline{\mathcal{C}}$.

Theorem 2.1.1 *If* $\overline{C} = \overline{F}\,\overline{X}$ *for some object* $\overline{X}$ *of* $\overline{\mathcal{X}}$, *then* $H_B^n(\overline{C}, \overline{G}) = 0$ *for all* $n \geq 2$. *Moreover,* $H_B^1(\overline{V}\,\overline{C}, \overline{G}) \cong \operatorname{Hom}_{\overline{C}}(\overline{V}\,\overline{C}, \overline{G})$ *for every object* $\overline{C}$ *of* $\overline{\mathcal{C}}$.

A morphism $\sigma \colon \overline{G} \longrightarrow \overline{G}'$ of abelian group objects induces for each $\overline{C}$ and n a homomorphism $\sigma^* \colon \operatorname{Hom}_{\overline{C}}(\overline{V}^n\overline{C}, \overline{G}) \longrightarrow \operatorname{Hom}_{\overline{C}}(\overline{V}^n\overline{C}, \overline{G}')$ that sends $\varphi \colon \overline{V}^n\overline{C} \longrightarrow \overline{G}$ to $\sigma \circ \varphi \colon \overline{V}^n\overline{C} \longrightarrow \overline{G}'$. In turn σ^* induces homomorphisms $H_B^n(\overline{C}, \overline{G}) \longrightarrow H_B^n(\overline{C}, \overline{G}')$.

A sequence $0 \longrightarrow \overline{G} \longrightarrow \overline{G}' \longrightarrow \overline{G}'' \longrightarrow 0$ of abelian group objects of $\overline{C}$ is *short V-exact* if and only if

$$0 \longrightarrow \operatorname{Hom}_{\overline{C}}(\overline{V}\overline{C}, \overline{G}) \longrightarrow \operatorname{Hom}_{\overline{C}}(\overline{V}\overline{C}, \overline{G}') \longrightarrow \operatorname{Hom}_{\overline{C}}(\overline{V}\overline{C}, \overline{G}'') \longrightarrow 0$$

is a short exact sequence of abelian groups for every object $\overline{C}$ of $\overline{C}$.

Theorem 2.1.2 *Every short V-exact sequence* $\mathcal{S} : 0 \longrightarrow \overline{G} \longrightarrow \overline{G}' \longrightarrow \overline{G}'' \longrightarrow 0$ *of abelian group objects of* $\overline{C}$ *induces an exact sequence*

$$\cdots H_B^n(\overline{C}, \overline{G}) \longrightarrow H_B^n(\overline{C}, \overline{G}') \longrightarrow H_B^n(\overline{C}, \overline{G}'') \longrightarrow H_B^{n+1}(\overline{C}, \overline{G}) \cdots$$

which is natural in $\mathcal{S}$.

Theorems 2.1.1 and 2.1.2 constitute Theorem 2 of [6]. These two properties determine H_B^n up to natural isomorphisms [5, Theorem 3.3]. Proposition 11.2 of [5] gives another characterization; see also [54]. For various reasons these characterizations do not apply directly to the cohomology theories in Sections 3 and 4.

2.1.6 Beck Extensions

The last major property of Beck cohomology is that H_B^2 classifies extensions.

A *Beck extension* [6] of an abelian group object $\overline{G}$ of $\overline{C}$ by the object S is an object $\overline{E} = (E, \sigma)$ of $\overline{C}$ such that every $\operatorname{Hom}_{\overline{C}}(\overline{C}, \overline{E})$ is a set and $U\sigma$ is a split epimorphism of $\mathcal{X}$ ($U\sigma \circ \nu = 1_{US}$ for some $\nu \colon US \longrightarrow UE$), together with, for each object $\overline{C} = (C, \pi)$ of $\overline{C}$, a simply transitive action $\cdot$ of $\operatorname{Hom}_{\overline{C}}(\overline{C}, \overline{G})$ on $\operatorname{Hom}_{\overline{C}}(\overline{C}, \overline{E})$ (for each $e, f \colon \overline{C} \longrightarrow \overline{E}$ there is a unique $g \colon \overline{C} \longrightarrow \overline{G}$ such that

$g \cdot e = f$), which preserves projection to S ($\sigma \circ (g \cdot e) = \sigma \circ e$ for all $g : \overline{C} \longrightarrow \overline{G}$ and $e : \overline{C} \longrightarrow \overline{E}$).

A morphism from $\overline{E} = (E, \sigma)$ to $\overline{F} = (F, \tau)$ of Beck extensions of $\overline{G}$ by S is a morphism $\varphi : \overline{E} \longrightarrow \overline{F}$ of $\overline{C}$ that preserves the action of $\overline{G}$ ($\varphi \circ (g \cdot e) = g \cdot (\varphi \circ e)$ for all $g : \overline{C} \longrightarrow \overline{G}$ and $e : \overline{C} \longrightarrow \overline{E}$).

Theorem 2.1.3 *If C is monadic over $\mathcal{X}$, then $H_B^2(S, \overline{G})$ classifies Beck extensions of $\overline{G}$ by S: there is a one-to-one correspondence, which is natural in $\overline{G}$, between the elements of $H_B^2(S, \overline{G})$ and the isomorphy classes of Beck extensions of $\overline{G}$ by S.*

This is Theorem 6 of [6].

2.2   Commutative Semigroups

We now let S be a given commutative semigroup, let C be the category of commutative semigroups, and let $\mathcal{X}$ be the category of sets.

2.2.1   Commutative Semigroups Over S

When $\overline{C} = (C, \pi)$ is a commutative semigroup over S, the homomorphism $\pi : C \longrightarrow S$ induces on C sets

$$C_a = \{ c \in C \mid \pi c = a \},$$

which may be empty, such that $C_a C_b \subseteq C_{ab}$ and C is the disjoint union $C = \bigcup_{a \in S} C_a$.

The category $\overline{C}$ of commutative semigroups over S inherits products from C. Given commutative semigroups $\overline{A} = (A, \pi)$ and $\overline{B} = (B, \rho)$ over S, construct the pullback in C:

$$
\begin{array}{ccc}
C & \xrightarrow{\pi'} & A \\
{\scriptstyle \rho'}\downarrow & & \downarrow{\scriptstyle \pi} \\
B & \xrightarrow{\rho} & S
\end{array}
$$

where

$$C = \{ (x, y) \in A \times B \mid \pi x = \rho y \},$$

$\pi'(x, y) = x$, and $\rho'(x, y) = y$. Then $\overline{C} = (C, \sigma) \in \overline{\mathcal{C}}$, where $\sigma = \pi \circ \pi' = \rho \circ \rho'$, is the product $\overline{A} \times \overline{B}$ in $\overline{\mathcal{C}}$, with projections π', ρ' to $\overline{A}$ and $\overline{B}$.

We use a *probe* to investigate individual commutative semigroups over S. For each $a \in S$ let $\overline{P_a} = (P, \psi_a)$, where $P = \{p, p^2, \ldots, p^n, \ldots\}$ and $\psi_a \colon P \longrightarrow S$ is the homomorphism that sends p to a, and p^n to a^n. If $\overline{C} = (C, \pi)$ is a commutative semigroup over S, then there is for each $c \in C$ a unique homomorphism $\psi_c \colon \overline{P_a} \longrightarrow \overline{C}$ such that $\psi_c(p) = c$, where $a = \pi c$. This provides a *probe bijection*

$$\mathrm{Hom}_{\overline{\mathcal{C}}}(\overline{P_a}, \overline{C}) \longrightarrow C_a = \{c \in C \mid \pi c = a\}, \quad \varphi \longmapsto \varphi(p), \tag{2.10}$$

where $\varphi(p) \in C_a$ since φ preserves projection to S.

2.2.2 Abelian Group Objects Over S

Probing an abelian group object $\overline{G} = (G, \rho)$ of $\overline{\mathcal{C}} = \mathcal{C} \downarrow S$ yields an abelian group valued functor $\mathbb{A}(\overline{G})$. For each $a \in S$ the probe bijection (2.10) of $\mathrm{Hom}_{\overline{\mathcal{C}}}(\overline{P_a}, \overline{G})$ onto $G_a = \{x \in G \mid \rho x = a\}$ carries the addition on $\mathrm{Hom}_{\overline{\mathcal{C}}}(\overline{P_a}, \overline{G})$ to an addition on G_a, namely

$$x + y = (\psi_x + \psi_y)(p),$$

where $\psi_x \colon \overline{P_x} \longrightarrow \overline{G}$ is the homomorphism that sends p to x, and similarly for ψ_y. This makes G_a an abelian group. In particular, $G_a \neq \varnothing$. Let z_a denote the identity element of G_a. If the commutative semigroup G lacks an identity element, then let $z_1 = 1 \in G^1$ and $G_1 = \{z_1\}$, so that $z_1 z_a = z_a$ for all $a \in S^1$.

The new additions cohabitate gracefully with the multiplication on G:

Lemma 2.2.1 *For all $a, b \in S$, $g, x \in G_a$, and $h, y \in G_b$,*

$$(g + x)(h + y) = gh + xy$$

In particular, $z_a z_b = z_{ab}$.

Proof This is Lemma VII.2.4 of [31]. Construct $\overline{H} = \overline{G} \times \overline{G}$ and its projections ρ', ρ'' to $\overline{G}$, from the pullback

$$\begin{array}{ccc} H & \xrightarrow{\rho'} & G \\ {\scriptstyle \rho''}\downarrow & & \downarrow{\scriptstyle \rho} \\ G & \xrightarrow{\rho} & S, \end{array}$$

so that $\overline{H} = (H, \sigma)$, where

$$H = \{(x, y) \in G \times G \mid \rho x = \rho y\},$$

$\sigma = \rho \circ \rho' = \rho \circ \rho''$, $\rho'(x, y) = x$, and $\rho''(x, y) = y$. The map

$$\alpha = \rho' + \rho'' \in \mathrm{Hom}_{\overline{C}}(\overline{H}, \overline{G})$$

determines the addition on every $\mathrm{Hom}_{\overline{C}}(\overline{C}, \overline{G})$ as well as the partial addition on G, as follows. If $\overline{C}$ is a commutative semigroup over S, then $g, h \in \mathrm{Hom}_{\overline{C}}(\overline{C}, \overline{G})$ induce $(g, h) \in \mathrm{Hom}_{\overline{C}}(\overline{C}, \overline{G} \times \overline{G})$ such that $\rho' \circ (g, h) = g$ and $\rho'' \circ (g, h) = h$, namely $(g, h)(c) = (g(c), h(c))$ for all $c \in C$; then

$$g + h = \rho' \circ (g, h) + \rho'' \circ (g, h) = (\rho' + \rho'') \circ (g, h) = \alpha \circ (g, h).$$

In particular, $(g+h)(c) = \alpha(g(c), h(c))$ for all $c \in C$; if $x, y \in G_a$, then $(x, y) \in H$ and

$$x + y = (\psi_x + \psi_y)(p) = \alpha(\psi_x(p) + \psi_y(p)) = \alpha(x, y).$$

If now $g, x \in G_a$ and $h, y \in G_b$, then

$$(g + x)(h + y) = \alpha(g, x)\alpha(h, y) = \alpha\big((g, x)(h, y)\big) = \alpha(gh, xy) = gh + xy,$$

since α is a multiplicative homomorphism. In particular,

$$z_a z_b = (z_a + z_a)(z_b + z_b) = z_a z_b + z_a z_b,$$

and $z_a z_b = z_{ab}$, since $z_a z_b \in G_{ab}$.   □

It follows from Lemma 2.2.1 that

$$(x + y)z_t = (x + y)(z_t + z_t) = x z_t + y z_t$$

whenever $x, y \in G_a$ and $t \in S^1$, so that

$$\gamma_{a,t} : g \longmapsto g z_t : G_a \longrightarrow G_{at} \tag{2.11}$$

is a homomorphism of G_a into G_{at}. Moreover, $(g z_t) z_u = g z_{tu}$. Hence $\mathbb{A}(\overline{G}) = (G, \gamma)$ is an abelian group valued functor on S.

If we let G_a act on itself by addition, $g \cdot h = g + h$, then $(G, \rho, \cdot)$ is a split coextension of S by $\mathbb{A}(\overline{G})$.

Conversely, given an abelian group valued functor $\mathcal{G} = (G, \gamma)$ on S, an abelian group object $\mathbb{O}(\mathcal{G})$ can be constructed as follows. Let $(E, \rho, \cdot) = \mathrm{E}(\mathcal{G})$ be the canonical split coextension of S by $\mathcal{G}$ from Chap. 1, so that

$$E = \{(g,a) \mid a \in S, g \in G_a\},$$

$(g,a)(h,b) = (g^b + h^a, ab)$, $\rho(g,a) = a$, and $g \cdot (h,a) = (g+h, a)$, for all a, b, g, h. Let $\overline{E} = (E, \rho)$. For every commutative semigroup $\overline{C} = (C, \pi)$ over S, define an addition on $\mathrm{Hom}_{\overline{C}}(\overline{C}, \overline{E})$ as follows: if $\varphi, \psi : \overline{C} \longrightarrow \overline{E}$ are homomorphisms and $c \in C$, $a = \pi c$, then $\varphi(c) = (g_c, a)$ and $\psi(c) = (h_c, a)$ for some $g_c, h_c \in G_a$; let $(\varphi + \psi)(c) = (g_c + h_c, a)$. With this addition $\overline{E}$ becomes an abelian group object $\mathbb{O}(\mathcal{G})$ of $\overline{C}$.

The constructions $\mathbb{A}$ and $\mathbb{O}$ will be used hereafter. It is readily verified that they constitute an equivalence of categories. With a slight abuse of language we obtain:

Theorem 2.2.1 *An abelian group object of $C \downarrow S$ is a split coextension of S, and the category of abelian group objects of $C \downarrow S$ is equivalent to the category of abelian group valued functors on S.*

Theorem 2.2.1 is Proposition XII.2.1 of [31], from [21] and [25].

2.2.3 Beck Extensions of S

Beck extensions in $C \downarrow S$ are analyzed like its abelian group objects and turn out to be familiar objects: commutative group coextensions of S.

Let $\overline{E} = (E, \sigma)$ be a Beck extension of an abelian group object $\overline{G} = (G, \rho)$ by S. For each $a \in S$ the probe bijections (2.10)

$$\mathrm{Hom}_{\overline{C}}(\overline{P}_a, \overline{E}) \longleftrightarrow E_a, \quad \mathrm{Hom}_{\overline{C}}(\overline{P}_a, \overline{G}) \longleftrightarrow G_a$$

carry the simply transitive action of $\mathrm{Hom}_{\overline{C}}(\overline{P}_a, \overline{G})$ on $\mathrm{Hom}_{\overline{C}}(\overline{P}_a, \overline{E})$ to a simply transitive action $\cdot$ of G_a on E_a.

Lemma 2.2.2 *For all $a, b \in S$, $g \in G_a$, $h \in G_b$, $x \in E_a$, and $y \in E_b$,*

$$(g \cdot x)(h \cdot y) = gh \cdot xy. \tag{2.12}$$

This is Lemma XII.3.3 of [31]. It is proved like Lemma 2.2.1, using the direct product $\overline{G} \times \overline{E}$ in $\overline{C}$.

If z_a is the identity element of G_a, then (2.12) yields

$$(g \cdot x)\, y = (g \cdot x)(z_b \cdot y) = g z_b \cdot xy = \gamma_{a,b} g \cdot xy$$

for all $a, b \in S$, $g \in G_a$, $x \in E_a$, and $y \in E_b$, where we recognize in $\gamma_{a,b}$ the map $G_a \longrightarrow G_{ab}$ in $\mathbb{A}(\overline{G})$. Thus $\mathbb{C}(\overline{E}) = (E, \sigma, \cdot)$ is a commutative group coextension of S by $\mathbb{A}(\overline{G})$.

Conversely, given a commutative group coextension $E = (E, \sigma, \cdot)$ of S by an abelian group valued functor $\mathcal{G} = (G, \gamma)$ on S, construct a Beck extension $\mathbb{B}(E)$ as follows. First, $\overline{E} = (E, \sigma)$ is a commutative semigroup over S, and the surjection σ is a split epimorphism in the category of sets. Let $\mathbb{E}(\mathcal{G}) = (G, \rho, \cdot)$ be the canonical split coextension, so that $\mathbb{O}(\mathcal{G}) = \overline{G} = (G, \rho)$. Use the action of each G_a on E_a to define an action of $\mathrm{Hom}_{\overline{C}}(\overline{C}, \overline{G})$ on $\mathrm{Hom}_{\overline{C}}(\overline{C}, \overline{E})$ for each commutative semigroup $\overline{C} = (C, \pi)$ over S: if $c \in C$ and $g: \overline{C} \longrightarrow \overline{G}, u: \overline{C} \longrightarrow \overline{E}$ are homomorphisms, then

$$(g \cdot u)(c) = g(c) \cdot u(c)$$

With this action, $\overline{E}$ becomes a Beck extension $\mathbb{B}(E)$ of $\overline{G} = \mathbb{O}(\mathcal{G})$ by S.

It is readily verified that $\mathbb{C}(\mathbb{B}(E))$ is equivalent to E and that $\mathbb{B}(\mathbb{C}(\overline{E})) \cong \overline{E}$, for every E and $\overline{E}$. With another slight abuse of language we obtain Proposition XII.3.1 of [31], from [21] and [25]:

Theorem 2.2.2 *Let $\overline{G}$ be an abelian group object of $\overline{C} = C \downarrow S$. A Beck extension of $\overline{G}$ by S is a commutative group coextension of S by $\mathbb{A}(\overline{G})$. The category of Beck extensions of $\overline{G}$ by S is equivalent to the category of commutative group coextensions of S by $\mathbb{A}(\overline{G})$.*

2.3   Beck Cohomology of Commutative Semigroups

We now unravel the definitions in Sects. 2.1 and 2.2 to obtain a more concrete definition of the Beck cohomology of commutative semigroups.

2.3.1   The 'Free Commutative Semigroup' Adjunction

The category $\mathcal{C}$ of commutative semigroups has a typical adjunction from the category $\mathcal{X}$ of sets, which we denote by $(\mathbb{F}, \mathbb{U}, \eta, \varepsilon)$.

The functor $\mathbb{U}$ assigns to each commutative semigroup S its underlying set, also denoted by S.

The functor $\mathbb{F}$ assigns to each set X the free commutative semigroup $\mathbb{F}X$ on X. The elements of $\mathbb{F}X$ are normally written as nonempty commutative words $x_1 x_2 \cdots x_n$, where $n \geq 1$ and $x_1, x_2, \ldots, x_n \in X$. (Allowing the empty word yields the free commutative monoid on X.) To avoid confusion with products in S we write the elements of $\mathbb{F}X$ as nonempty *unordered* sequences $[x_1, x_2, \ldots, x_n]$ of elements of X; *unordered* means

$$[x_{\sigma 1}, x_{\sigma 2}, \ldots, x_{\sigma n}] = [x_1, x_2, \ldots, x_n]$$

for every permutation σ of $1, 2, \ldots, n$. Multiplication in $\mathbb{F}X$ is by concatenation:

$$[x_1, \ldots, x_m][y_1, \ldots, y_n] = [x_1, \ldots, x_m, y_1, \ldots, y_n]$$

The natural mapping $\eta_X \colon X \longrightarrow \mathbb{U}\mathbb{F}X$ sends $x \in X$ to the one-letter word $[x]$.

For every mapping f of X into a commutative semigroup S there is a unique homomorphism φ of $\mathbb{F}X$ into X such that $\mathbb{U}\varphi \circ \eta_X = f$, equivalently $\varphi[x] = f(x)$ for all x; then φ sends $[x_1, x_2, \ldots, x_n] \in \mathbb{F}X$ to the product $f(x_1)f(x_2)\cdots f(x_n)$ in S.

Similarly every mapping $f \colon X \longrightarrow Y$ induces a unique homomorphism $\mathbb{F}f \colon \mathbb{F}X \longrightarrow \mathbb{F}Y$ such that $\mathbb{U}\mathbb{F}f \circ \eta_X = \eta_Y \circ f$; namely,

$$(\mathbb{F}f)[x_1, x_2, \ldots, x_n] = [f(x_1), f(x_2), \ldots, f(x_n)] \tag{2.13}$$

The natural transformation ε may be less familiar. The general property (9.34): $\mathbb{U}\varepsilon_S \circ \eta_{\mathbb{U}S} = 1_{\mathbb{U}S}$ implies that ε_S is the unique homomorphism of $\mathbb{F}\mathbb{U}S$ into S that sends $[a]$ to a for every $a \in S$. Hence ε_S sends a commutative word $[a_1, a_2, \ldots, a_n] \in \mathbb{F}\mathbb{U}S$ to its product $a_1 a_2 \cdots a_n$ in S:

$$\varepsilon_S[a_1, a_2, \ldots, a_n] = a_1 a_2 \cdots a_n \tag{2.14}$$

2.3.2   The 'Free Commutative Semigroup' Comonad

The adjunction $(\mathbb{F}, \mathbb{U}, \eta, \varepsilon)$ from $\mathcal{X}$ to $\mathcal{C}$ begets a comonad $(\mathbb{V}, \varepsilon, \nu)$ on $\mathcal{C}$, in which $\mathbb{V} = \mathbb{F}\mathbb{U} \colon \mathcal{C} \longrightarrow \mathcal{C}$ and $\nu_C = \mathbb{F}\eta_{\mathbb{U}C} \colon \mathbb{V}C \longrightarrow \mathbb{V}\mathbb{V}C$, for every commutative semigroup C.

Thus $\mathbb{V} = \mathbb{F}\mathbb{U}$ sends a commutative semigroup C to the free commutative semigroup $\mathbb{F}\mathbb{U}C$ on the set C. If $\varphi \colon C \longrightarrow C'$ is a homomorphism, then, by (2.13),

$$(\mathbb{V}\varphi)[c_1, c_2, \ldots, c_n] = [\varphi(c_1), \varphi(c_2), \ldots, \varphi(c_n)] \tag{2.15}$$

for every $[c_1, c_2, \ldots, c_n] \in \mathbb{V}C$. By (2.14), $\varepsilon_C \colon \mathbb{V}C \longrightarrow C$ sends an element $[c_1, c_2, \ldots, c_n]$ of $\mathbb{V}C$ to its product in C:

$$\varepsilon_C[c_1, c_2, \ldots, c_n] = c_1 c_2 \cdots c_n.$$

Finally, ν_C is the homomorphism induced by the mapping $\eta_{\mathbb{U}C}$ of $\mathbb{U}C$ into $\mathbb{F}\mathbb{U}C$, so that

$$\nu[c_1, c_2, \ldots, c_n] = [\eta(c_1), \eta(c_1), \ldots, \eta(c_1)] \tag{2.16}$$

$$= [[c_1], [c_2], \ldots, [c_n]].$$

for all $c_1, c_2, \ldots, c_n \in C$.

Now let S be a given commutative semigroup. The functor $\mathbb{V}$ lifts to a functor $\overline{\mathbb{V}} \colon \overline{\mathcal{C}} \longrightarrow \overline{\mathcal{C}}$. By (2.6),

$$\overline{\mathbb{V}}(T, \tau) = (\mathbb{V}T, \overline{\tau}), \text{ where } \overline{\tau} = \varepsilon_S \circ \mathbb{V}\tau$$

for every commutative semigroup (T, τ) over S. The typical element of $\mathbb{V}T = \mathbb{F}UT$ is a nonempty unordered sequence $[t_1, t_2, \ldots, t_n]$ of elements of T. By (2.13) and (2.14),

$$\overline{\tau}[t_1, t_2, \ldots, t_n] = \varepsilon_S[\tau t_1, \tau t_2, \ldots, \tau t_n] = \tau(t_1 t_2 \cdots t_n) \qquad (2.17)$$

for every $t_1, t_2, \ldots, t_n \in T$, since τ is a homomorphism.

2.3.3   Cochains

The cochains in Beck cohomology are homomorphisms $\overline{\mathbb{V}T} \longrightarrow \overline{G}$ and are unraveled as follows.

Given a commutative semigroup $\overline{T} = (T, \tau)$ over S and an abelian group valued functor $\mathcal{G}$ on S, a *cochain on $\overline{T}$ with values in $\mathcal{G}$* assigns to each $x \in T$ an element $u(x)$ of $G_{\tau x}$. Under pointwise addition, cochains on $\overline{T}$ with values in $\mathcal{G}$ constitute an abelian group $C\,(\overline{T}, \mathcal{G})$.

Lemma 2.3.1 *For every commutative semigroup $\overline{T}$ over S and abelian group object $\overline{G}$ of $\mathcal{C} \downarrow S$ there is an isomorphism*

$$\mathrm{Hom}_{\overline{\mathcal{C}}}\,(\overline{\mathbb{V}T}, \overline{G}) \cong C\,(\overline{T}, \mathbb{A}(\overline{G}))$$

which is natural in $\overline{T}$ and $\overline{G}$. The homomorphism $\varphi \colon \overline{\mathbb{V}T} \longrightarrow \overline{G}$ that corresponds to $u \in C\,(\overline{T}, \mathbb{A}(\overline{G}))$ sends $[t_1, \ldots, t_n] \in \mathbb{V}T$ to

$$\varphi\,[t_1, t_2, \ldots, t_n] = \Big(\sum_{1 \le i \le n} u(t_i)^{a_i^*}, a \Big), \qquad (2.18)$$

where $a_i = \tau t_i$, $a = a_1 \cdots a_n$, and $a_i^* = a_1 \cdots a_{i-1} a_{i+1} \cdots a_n$ is as in Lemma 1.2.1.

This is Lemma XII.4.2 of [31].

Proof Let $\overline{T} = (T, \tau)$. Since $\mathbb{V}T = \mathbb{F}UT$ is free on T, a homomorphism of $\overline{\mathbb{V}T}$ is uniquely determined by its values on T. By Theorem 2.2.1 we may assume that $\overline{G}$ is the canonical split coextension $\mathbb{E}(\mathcal{G}) = (E, \zeta)$. If $\varphi \colon \overline{\mathbb{V}}(T, \tau) \longrightarrow \overline{G}$ is a morphism in $\mathcal{C} \downarrow S$, then $\zeta \circ \varphi = \overline{\tau}$ and there is for each $x \in T$ some $u(x) \in G_{\overline{\tau}[x]} = G_{\tau x}$ such that $\varphi[x] = (u(x), \tau x)$. Then φ sends $[t_1, t_2, \ldots, t_n] = [t_1][t_2] \cdots [t_n] \in \mathbb{V}T$ to

$$\varphi\,[t_1, t_2, \ldots, t_n] = (u(t_1), a_1)(u(t_2), a_2) \ldots (u(t_n), a_n)$$

where $a_i = \tau t_i$. By Lemma 1.2.1, recast in the additive notation,

$$(u(t_1), a_1)(u(t_2), a_2) \ldots (u(t_n), a_n) = \left(\sum_{1 \le i \le n} u(t_i)^{a_i^*}, a \right).$$

Conversely, since $\mathbb{V}T$ is free on T, there is for each $u \in C(\overline{T}, \mathcal{G})$ a unique semigroup homomorphism $\varphi \colon \mathbb{V}T \longrightarrow E$ such that $\varphi[x] = (u(x), \tau x)$ for all $x \in T$. This yields the inverse isomorphism $C(\overline{T}, \mathbb{A}(\overline{G})) \longrightarrow \text{Hom}_{\overline{C}}(\overline{\mathbb{V}T}, \overline{G})$. $\qquad \square$

By Lemma 2.3.1, the complex (2.7) whose homology groups are the Beck cohomology groups of S:

$$0 \longrightarrow \text{Hom}_{\overline{C}}(\overline{\mathbb{V}S}, \overline{G}) \ldots \text{Hom}_{\overline{C}}(\overline{\mathbb{V}^n S}, \overline{G}) \xrightarrow{\delta_n} \text{Hom}_{\overline{C}}(\overline{\mathbb{V}^{n+1} S}, \overline{G}) \ldots$$

$$(2.19)$$

where $\overline{S} = (S, 1_S)$, is isomorphic to

$$0 \longrightarrow C(\overline{S}, \mathcal{G}) \longrightarrow \ldots \longrightarrow C(\overline{T}_{n-1}, \mathcal{G}) \xrightarrow{\delta_n} C(\overline{T}_n, \mathcal{G}) \longrightarrow \ldots$$

where $\overline{T}_n = \overline{\mathbb{V}^n S}$ if $n > 0$, $\overline{T}_0 = \overline{S}$ and $\delta_n \colon C(\overline{T}_{n-1}, \mathcal{G}) \longrightarrow C(\overline{T}_n, \mathcal{G})$ is induced by $\delta_n \colon \text{Hom}_{\overline{C}}(\overline{\mathbb{V}^n S}, \overline{G}) \longrightarrow \text{Hom}_{\overline{C}}(\overline{\mathbb{V}^{n+1} S}, \overline{G})$ and described in more detail below.

For $n \ge 1$ the elements of

$$C_B^n(S, \mathcal{G}) = C(\overline{T}_{n-1}, \mathcal{G})$$

are *Beck n-cochains* on S with values in $\mathcal{G}$.

Let $\overline{T}_n = (T_n, \pi_n) = \overline{\mathbb{V}^n S}$. Then $T_0 = S, \pi_0 = 1_S$; $T_n = \mathbb{V}^n S$; $T_{n+1} = \mathbb{V}T_n$ is the semigroup of all nonempty unordered sequences $[x_1, x_2, \ldots, x_m]$ of elements of T_n; and $\pi_{n+1} = \overline{\pi}_n$, so that (2.17) yields

$$\pi_{n+1}[x_1, x_2, \ldots, x_m] = \pi_n (x_1 x_2 \cdots x_m)$$

for all $m > 0$ and $x_1, x_2, \ldots, x_m \in T_n$.

Thus, a Beck 1-cochain u on S with values in $\mathcal{G}$ assigns to each $a \in S$ an element $u(a)$ of G_a. If $n \ge 2$, then a Beck n-cochain u on S with values in $\mathcal{G}$ assigns to each $[x_1, \ldots, x_m] \in T_{n-1}$ an element $u(x_1, \ldots, x_m)$ of G_a, where $x_1, \ldots, x_m \in T_{n-2}$ and $a = \pi_{n-1}[x_1, \ldots, x_m] = \pi_{n-2}(x_1 \cdots x_m)$.

2.3.4 Cohomology

We saw that the coboundary $\delta_n : \mathrm{Hom}_{\overline{C}} (\overline{\mathbb{V}}^n \overline{S}, \overline{G}) \longrightarrow \mathrm{Hom}_{\overline{C}} (\overline{\mathbb{V}}^{n+1} \overline{S}, \overline{G})$ in (2.7) assigns

$$\delta_n(\varphi) \;=\; \sum_{0 \le i \le n} (-1)^i \, (\varphi \circ \overline{\varepsilon}_S^{\,n,i})$$

to $\varphi : \overline{\mathbb{V}}^n \overline{S} \longrightarrow \overline{G}$, as in (2.8), where

$$\overline{\varepsilon}^{\,n,i} \;=\; \varepsilon^{n,i} \;=\; \mathbb{V}^{n-i} \, \overline{\varepsilon} \, \mathbb{V}^i : \mathbb{V}^{n+1} \longrightarrow \mathbb{V}^n$$

so that

$$\overline{\varepsilon}_{\overline{T}}^{\,n,i} \;=\; \varepsilon_{\overline{T}}^{\,n,i} \;=\; \mathbb{V}^{n-i} \, \varepsilon_{\mathbb{V}^i T} : \mathbb{V}^{n+1} T \longrightarrow \mathbb{V}^n T$$

for every $\overline{T} = (T, \tau)$, in particular

$$\varepsilon_S^{\,n,i} \;=\; \mathbb{V}^{n-i} \, \varepsilon_{T_i} : T_{n+1} \longrightarrow T_n$$

The homomorphisms $\varepsilon^{n,i} = \varepsilon_S^{\,n,i}$ can be constructed recursively: since $\varepsilon_S^{\,n,n} = \varepsilon_{T_n}$, we have

$$\varepsilon^{n,n} [x_1, x_2, \dots, x_m] \;=\; x_1 x_2 \cdots x_m \tag{2.20}$$

for all $x_1, x_2, \dots, x_m \in T_n$; if $i < n$, then $\varepsilon^{n,i} = \mathbb{V}\varepsilon^{n-1,i}$ and

$$\varepsilon^{n,i} [x_1, x_2, \dots, x_m] \;=\; [\varepsilon^{n-1,i} x_1, \, \varepsilon^{n-1,i} x_2, \, \dots, \, \varepsilon^{n-1,i} x_m] \tag{2.21}$$

for all $x_1, x_2, \dots, x_m \in T_n$, by (2.15). In particular,

$$\varepsilon^{0,0} [a_1, a_2, \dots, a_m] \;=\; a_1 a_2 \cdots a_m \tag{2.22}$$

for all $a_1, a_2, \dots, a_m \in S$, and

$$\varepsilon^{1,1} [A_1, A_2, \dots, A_m] \;=\; A_1 A_2 \cdots A_m , \tag{2.23}$$

$$\varepsilon^{1,0} [A_1, A_2, \dots, A_m] \;=\; [a_1, a_2, \dots, a_m] \tag{2.24}$$

for all $A_1, A_2, \dots, A_m \in T_1$, where $a_i = \varepsilon^{0,0} A_i = \pi_1 A_i$.

The homomorphism $\delta_n : C_B^n (S, \mathcal{G}) \longrightarrow C_B^{n+1} (S, \mathcal{G})$ induced by the general coboundary $\delta_n : \mathrm{Hom}_{\overline{C}} (\overline{\mathbb{V}}^n \overline{S}, \overline{G}) \longrightarrow \mathrm{Hom}_{\overline{C}} (\overline{\mathbb{V}}^{n+1} \overline{S}, \overline{G})$ can now be made more explicit. Let $\mathcal{G} = \mathbb{A}(\overline{G})$. The isomorphism

$$C_B^n(S, \mathcal{G}) = C(\overline{T}_{n-1}, \mathcal{G}) \cong \mathrm{Hom}_{\overline{C}}(\overline{\nabla}^n S, \overline{G}) = \mathrm{Hom}_{\overline{C}}(\overline{T}_n, \overline{G})$$

in Lemma 2.3.1 sends $u \in C_B^n(S, \mathcal{G})$ to the homomorphism $\varphi \colon \overline{T}_n \longrightarrow \overline{G}$ such that $\varphi[t] = (u(t), \pi_{n-1}t)$ for all $t \in T_{n-1}$. Then δ_n sends φ to

$$\delta_n \varphi = \sum_{0 \leq i \leq n} (-1)^i \, (\varphi \circ \overline{\varepsilon}_S^{n,i}) \tag{2.25}$$

$$= \sum_{0 \leq i \leq n-1} (-1)^i \, (\varphi \circ \varepsilon^{n,i}) + (-1)^n (\varphi \circ \varepsilon^{n,n})$$

and the isomorphism $\mathrm{Hom}_{\overline{C}}(\overline{T}_{n+1}, \overline{G}) \cong C_B^{n+1}(S, \mathcal{G})$ in Lemma 2.3.1 sends $\delta_n \varphi$ to a cochain $v \in C_B^{n+1}(S, \mathcal{G})$; if $t \in T_n$, then

$$v(t) = (\delta_n \varphi)[t] = \sum_{0 \leq i \leq n-1} (-1)^i \, (\varphi(\varepsilon^{n,i}[t])) + (-1)^n (\varphi(\varepsilon^{n,n}[t]))$$

$$= \sum_{0 \leq i \leq n-1} (-1)^i \, (\varphi[\varepsilon^{n-1,i}t]) + (-1)^n (\varphi(\varepsilon^{n,n}[t]))$$

If $t = [t_1, t_2, \ldots, t_m]$, where $t_1, t_2, \ldots, t_m \in T_{n-1}$, and $\pi_{n-1}t_i = a_i$ then $\varphi t_i = (u(t_i, a_i)$ and (2.20), (2.18), (2.21) yield

$$\varphi(\varepsilon^{n,n}[t]) = \varphi(t_1 t_2 \cdots t_m) = (\varphi t_1)(\varphi t_2) \cdots (\varphi t_m)$$

$$= (u(t_1), a_1)(u(t_2), a_2) \cdots (u(t_m), a_m)$$

$$= \Big(\sum_{1 \leq i \leq m} u(t_i)^{a_i^*}, a \Big)$$

where $a = a_1 a_2 \cdots a_m$ and $a_i^* = a_1 \cdots a_{i-1} a_{i+1} \cdots a_m$. Hence

$$(\delta_n u)(t) = \sum_{0 \leq i \leq n-1} (-1)^i \, (u(\varepsilon^{n-1,i}t)) + (-1)^n \sum_{1 \leq i \leq m} u(t_i)^{a_i^*} \tag{2.26}$$

for all $t = [t_1, t_2, \ldots, t_m] \in T_n$, by (2.25).

Theorem 2.3.1 *Up to natural isomorphisms the Beck cohomology of a commutative semigroup S has coefficients in an abelian group valued functor $\mathcal{G}$ on S, and $H_B^n(S, \mathcal{G})$ is the n-th homology group of the complex*

$$0 \longrightarrow C_B^1(S, \mathcal{G}) \longrightarrow \cdots \longrightarrow C_B^n(S, \mathcal{G}) \xrightarrow{\delta_n} C_B^{n+1}(S, \mathcal{G}) \longrightarrow \cdots$$

where δ_n is given by (2.26).

More detailed proofs of Theorem 2.3.1 can be found in [31] and [25].

Given an abelian group valued functor $\mathcal{G}$ on S, a *Beck n-cocycle* is an element of $Z_B^n(S, \mathcal{G}) = \operatorname{Ker} \delta_n$; a *Beck n-coboundary* is an element of $B_B^n(S, \mathcal{G}) = \operatorname{Im} \delta_{n-1}$. The inclusion $B_B^n(S, \mathcal{G}) \subseteq Z_B^n(S, \mathcal{G})$ follows from (2.4): $\delta_n \circ \delta_{n-1} = 0$. Then

$$H_B^n(S, \mathcal{G}) \cong Z_B^n(S, \mathcal{G}) \, / \, B_B^n(S, \mathcal{G})$$

If for example $t = [a_1, a_2, \ldots, a_m] \in T_1$, where $a_1, a_2, \ldots, a_m \in S$, and $u \in C_B^1(S, \mathcal{G})$, then

$$
\begin{aligned}
(\delta_1 u)(t) &= u(\varepsilon^{0,0} t) - \sum_{1 \le i \le m} u(a_i)^{a_i^*} \\
&= u(a_1 a_2 \cdots a_m) - \sum_{1 \le i \le m} u(a_i)^{a_i^*} \qquad (2.27)
\end{aligned}
$$

by (2.22). Hence $u \in C_B^1(S, \mathcal{G})$ is a Beck 1-cocycle if and only if

$$u(a_1 a_2 \cdots a_m) = \sum_{1 \le i \le m} u(a_i)^{a_i^*} \qquad (2.28)$$

for all $a_1, a_2, \ldots, a_m \in S$; equivalently,

$$u(ab) = u(a)^b + u(b)^a \qquad (2.29)$$

for all $a, b \in S$. Beck 1-cocycles are also called *derivations*, after the Calculus formula $d\,(hi\,ho) = (d\,hi)\,ho + hi\,d\,ho$.

If $t = [A_1, \ldots, A_m] \in T_2$, where $A_1, \ldots, A_m \in T_1$, and $u \in C_B^2(S, \mathcal{G})$, then

$$(\delta_2 u)(t) = u(\varepsilon^{1,0} t) - u(\varepsilon^{1,1} t) + \sum_{1 \le i \le m} u(a_i)^{a_i^*}$$

so that (2.23) and (2.24) yield

$$(\delta_2 u)[A_1, \ldots, A_m] = u([a_1, \ldots, a_m]) - u(A_1 \cdots A_m) + \sum_{1 \le i \le m} u(A_i)^{a_i^*}$$

for all $A_1, \ldots, A_m \in T_1$, where $a_i = \pi A_i$. Hence $u \in C_B^2(S, \mathcal{G})$ is a Beck 2-cocycle if and only if

$$u(A_1 A_2 \cdots A_m) = u([a_1, a_2, \ldots, a_m]) + \sum_{1 \le i \le m} u(A_i)^{a_i^*} \qquad (2.30)$$

for all $A_1, A_2, \ldots, A_m \in T_1$.

2.3.5 Properties

Properties of the Beck cohomology of commutative semigroups follow from general properties of Beck cohomology in Sect. 2.1. By Theorem 2.1.1:

Theorem 2.3.2 *If S is a free commutative semigroup, then $H_B^n(S, \mathcal{G}) = 0$ for all $n \geq 2$ and every abelian group valued functor $\mathcal{G}$ on S.*

Every short exact sequence $0 \longrightarrow \mathcal{G} \longrightarrow \mathcal{G}' \longrightarrow \mathcal{G}'' \longrightarrow 0$ of abelian group valued functors on S is short $\mathbb{V}$-exact. Hence Theorem 2.1.2 yields:

Theorem 2.3.3 *Every short exact sequence $\mathcal{S} :\longrightarrow \mathcal{G} \longrightarrow \mathcal{G}' \longrightarrow \mathcal{G}'' \longrightarrow 0$ of abelian group valued functors on S induces an exact sequence*

$$\cdots H_B^n(S, \mathcal{G}) \longrightarrow H_B^n(S, \mathcal{G}') \longrightarrow H_B^n(S, \mathcal{G}'') \longrightarrow H_B^{n+1}(S, \mathcal{G}) \cdots$$

which is natural in $\mathcal{S}$.

Theorem 2.3.3 also follows from the construction in Sect. XII.4 of [31] of a projective complex of abelian group valued functors on S, of which H_B^n is the cohomology.

Finally, combining Theorems 2.1.3 and 2.2.2 yields:

Theorem 2.3.4 *$H_B^2(S, \mathcal{G})$ classifies commutative group coextensions of S by $\mathcal{G}$: there is one-to-one correspondence, which is natural in $\mathcal{G}$, between the elements of $H_B^2(S, \mathcal{G})$ and the equivalence classes of commutative group coextensions of S by $\mathcal{G}$.*

The less than obvious relationship between H_B^2 and commutative group coextensions will be clarified in the next chapter by a more direct proof of Theorem 2.3.4.

One property that Beck cohomology does *not* have is computability: if S and all groups G_a are finite, then $C_B^1(S, \mathcal{G})$ is finite, but if $n \geq 2$ then the variables in Beck n-cochains are elements of the infinite semigroup T_n and $C_B^n(S, \mathcal{G})$ is infinite as well. Thus only $H_B^1(S, \mathcal{G})$ is computable.

Chapter 3
Symmetric Cohomology

The Beck cohomology of commutative semigroups S has impractical cochains. The variables of Beck 2-cochains on S, for example, are unordered sequences of unbounded length of unordered sequences of unbounded length of elements of S; this makes finding $H_B^2(S, \mathcal{G})$ an infinite task, even if S and all groups G_a are finite. Also, apart from Theorem 2.3.4, H_B^2 lacks a direct connection to the coextensions in Chap. 1.

The same can be said of the Beck cohomology of groups, in which the variables of 2-cochains on a group G are reduced words of unbounded length, the letters of which are reduced words of unbounded length, the letters of which are elements of G. The cohomology of groups, however, has an equivalent and earlier definition [9, 41], in which n-cochains are functions of n variables in G.

It is clear from Chap. 1 that $H^2(S, \mathcal{G})$ has an equivalent, and more natural, definition, in which 2-cochains are functions of two variables in S. Unfortunately, extending this definition to higher dimensions runs into difficulties (addressed in Sect. 3.2 and in Chap. 8) that have, as of this writing, been overcome only in dimensions 3 and 4.

This chapter states the more natural definition, compares the new cohomology to Beck cohomology, and states its main properties.

3.1 Definition

The definition of symmetric cohomology follows the hints in Sect. 1.2.

3.1.1 Cochains

Let S be a commutative semigroup and let $\mathcal{G}$ be an abelian group valued functor on S.

An *n-cochain* u on S with values in $\mathcal{G}$, where $n \geq 1$, assigns to each $a_1, a_2, \ldots, a_n \in S$ an element $u(a_1, a_2, \ldots, a_n)$ of G_a, where $a = a_1 a_2 \cdots a_n$.

Under pointwise addition, n-cochains on S with values in $\mathcal{G}$ constitute an abelian group $C^n(S, \mathcal{G})$. In particular, $C^1(S, \mathcal{G}) = C_B^1(S, \mathcal{G})$.

The coboundary of an n-cochain u is the $(n + 1)$-cochain δu defined by:

$$(\delta_n u)(a_1, a_2, \ldots, a_{n+1}) = u(a_2, \ldots, a_{n+1})^{a_1} \tag{3.1}$$
$$+ \sum_{1 \leq i \leq n} (-1)^i u(a_1, \ldots, a_i a_{i+1}, \ldots, a_{n+1})$$
$$+ (-1)^{n+1} u(a_1, a_2, \ldots, a_n)^{a_{n+1}}.$$

Very similar definitions are used in the Leech cohomology [40] and in the cohomology of groups [41] (see also Sect. 9.1).

It is readily verified that $\delta_{n+1} \circ \delta_n = 0$, so that

$$0 \longrightarrow C^1(S, \mathcal{G}) \longrightarrow \cdots \longrightarrow C^n(S, \mathcal{G}) \xrightarrow{\delta_n} C^{n+1}(S, \mathcal{G}) \longrightarrow \cdots \tag{3.2}$$

is a complex of abelian groups. The homology groups of this complex (defined without any symmetry conditions) are the Leech cohomology groups $H_L^n(S, \mathcal{G})$ of the commutative semigroup S with coefficients in $\mathcal{G}$ [40]. The second group $H_L^2(S, \mathcal{G})$ classifies *all* group coextensions of S by $\mathcal{G}$, including all noncommutative coextensions (see Sect. 9.1).

3.1.2 Symmetric Cochains

In order to classify commutative group coextensions only, we saw in Chap. 1 that 2-cocycles must satisfy the commutativity condition

$$u(b, a) = u(a, b) \tag{S2}$$

for all $a, b \in S$. The coboundaries of 2-cochains with this property have properties of their own, which must then be passed on to 3-cochains. This in turn creates properties of their coboundaries, which must then be passed on to 4-cochains. Hence the next definition. It would be more logical to call symmetric 2-cochains *commutative*, but this would not fit higher dimensions, in view of (S3) and (S4) below.

Let $n = 1, 2, 3, 4$. We call an n-cochain u on S with values in $\mathcal{G}$ *symmetric* if and only if it satisfies the following condition (Sn).

(S1) is vacuous.

(S2) is the condition

$$u(b, a) = u(a, b) \tag{S2}$$

for all $a, b \in S$, as above.

(S3) consists of two conditions: for all $a, b, c \in S$,

$$u(c, b, a) = -u(a, b, c) \text{ and} \tag{S3a}$$

$$u(a, b, c) = u(b, a, c) - u(b, c, a). \tag{S3b}$$

The original definition of (S3) in [28] (also in [31]) includes a third condition

$$u(a, b, c) + u(b, c, a) + u(c, a, b) = 0$$

which follows from (S3b). Condition (S3a) also follows from (S3b) but is included above to show analogy with dimensions 2 and 4.

(S4) consists of three conditions: for all $a, b, c, d \in S$,

$$u(d, c, b, a) = -u(a, b, c, d), \tag{S4a}$$

$$u(a, b, c, d) = u(b, a, c, d) - u(b, c, a, d) + u(b, c, d, a), \text{ and} \tag{S4b}$$

$$u(a, b, b, a) = 0. \tag{S4c}$$

The original conditions in [32] (also in [31]) are in a different order and include a fourth condition

$$u(a, b, c, d) - u(b, c, d, a) + u(c, d, a, b) - u(d, a, b, c) = 0$$

which, to the author's embarrassment, turned out to follows from (S4a) and (S4b) [34].

3.1.3 Symmetric Cohomology

Under pointwise addition, symmetric n-cochains on S with values in $\mathcal{G}$ constitute an abelian group $C_S^n(S, \mathcal{G}) \subseteq C^n(S, \mathcal{G})$.

As expected, the coboundary (3.1) has the following property:

Proposition 3.1.1 *If $u \in C_S^n(S, \mathcal{G})$ is a symmetric n-cochain, where $n \leq 3$, then δu is a symmetric $(n + 1)$-cochain.*

Hence the complex (3.2) yields a short complex

$$0 \longrightarrow C_S^1(S, \mathcal{G}) \overset{\delta}{\longrightarrow} C_S^2(S, \mathcal{G}) \longrightarrow \cdots \longrightarrow C_S^4(S, \mathcal{G}) \overset{\delta}{\longrightarrow} C^5(S, \mathcal{G}) \qquad (3.3)$$

of abelian groups. The symmetric cohomology groups of S with coefficients in $\mathcal{G}$ are the homology groups of this complex. Thus, if S is a commutative semigroup, $\mathcal{G}$ is an abelian group valued functor on S, and $n = 1, 2, 3, 4$, then a *symmetric n-cocycle* on S with coefficients in $\mathcal{G}$ is an element of

$$Z_S^n(S, \mathcal{G}) = \operatorname{Ker} \delta \subseteq C_S^n(S, \mathcal{G}),$$

a *symmetric n-coboundary* on S with coefficients in $\mathcal{G}$ is an element of

$$B_S^n(S, \mathcal{G}) = \operatorname{Im} \delta \subseteq Z_S^n(S, \mathcal{G}),$$

and the *n-th symmetric cohomology group* of S with coefficients in $\mathcal{G}$ is

$$H_S^n(S, \mathcal{G}) = Z_S^n(S, \mathcal{G}) \,/\, B_S^n(S, \mathcal{G}).$$

In particular, a symmetric 2-cocycle is a symmetric 2-cochain u such that

$$u(b, c)^a - u(ab, c) + u(a, bc) - u(a, b)^c = 0$$

for all $a, b, c \in S$, which we recognize as a commutative factor set from Chap. 1; a symmetric 2-coboundary is a symmetric 2-cochain u such that $u = \delta v$ for some 1-cochain v:

$$u(a, b) = v(b)^a - v(ab) + v(a)^b$$

for all $a, b \in S$, which we recognize as the commutative factor set of a split coextension. Hence Theorem 1.2.1 yields:

Theorem 3.1.1 $H_S^2(S, \mathcal{G})$ classifies *commutative group coextensions of S by $\mathcal{G}$: there is a one-to-one correspondence between the elements of $H_S^2(S, \mathcal{G})$ and the equivalence classes of commutative group coextensions of S by $\mathcal{G}$.*

Section 3.4 contains further properties of H_S^1 and H_S^2.

3.1.4 An Example

With Theorem 3.1.1 it is possible to determine all equivalence classes of commutative group coextensions of a commutative semigroup S by an abelian group valued functor $\mathcal{G}$. This is practicable only if S and the groups G_a have very few elements.

In this example, S and $\mathcal{G}$ are the groupfree semigroup and Schützenberger functor of Example 1.

Example 2 Let S be the groupfree commutative monoid $\{1, a, e, 0\}$ with multiplication table

$$
\begin{array}{c|cccc}
1 & a & e & 0 \\
a & 0 & 0 & 0 \\
e & 0 & 0 & 0 \\
0 & 0 & 0 & 0
\end{array}
$$

Let $\mathcal{G} = (G, \gamma)$ be the Schützenberger functor of Example 1, so that $G_1 = \{0, g\}$, $G_a = \{0, h\}$, where $2g = 2h = 0$, $G_e = G_0 = 0$, and γ consists of: the four identity homomorphisms; the isomorphism $\gamma_{1,a} \colon G_1 \cong G_a$ that sends g to h, so that $g^a = h$; and trivial homomorphisms.

There are four 1-cochains in $C_S^1(S, \mathcal{G})$:

u	$u(1)$	$u(a)$	$u(e)$	$u(0)$
u_1	0	0	0	0
u_2	g	0	0	0
u_3	0	h	0	0
u_4	g	h	0	0

A symmetric 2-cochain v is determined by its values $v(1, 1) \in G_1$ and $v(1, a) = v(a, 1) \in G_a$: indeed $v(x, y) = 0$ for all other (x, y) since $G_e = G_0 = 0$. Hence $C_S^2(S, \mathcal{G})$ has four elements. In particular, a 2-coboundary δu, where $u \in C_S^1(S, \mathcal{G})$, is determined by:

$$
(\delta u)(1, 1) = u(1)^1 - u(1) + u(1)^1 = u(1) \quad \text{and}
$$

$$
(\delta u)(a, 1) = (\delta u)(1, a) = u(a)^1 - u(a) + u(1)^a = u(1)^a,
$$

so that

δu	$(\delta u)(1, 1)$	$(\delta u)(1, a)$
δu_1	0	0
δu_2	g	h
δu_3	0	0
δu_4	g	h

and $B_S^2(S, \mathcal{G}) = \{\delta u_1, \delta u_2\}$ has two elements. The group $Z_S^1(S, \mathcal{G}) = \{u_1, u_3\}$ also has two elements.

By Proposition 3.1.1, if $v \in C_S^2(S, \mathcal{G})$, then δv has properties (S3a): $(\delta v)(z, y, x) = -(\delta v)(x, y, z)$ and (S3b), which implies $(\delta v)(x, y, z) = 0$ if $x = z$; since $(\delta v)(x, y, z) \in G_{xyz} = 0$ if $xyz = 0$ or $xyz = e$, it follows that δv is completely determined by

$$(\delta v)(1, 1, a) = v(1, a)^1 - v(11, a) + v(1, 1a) - v(1, 1)^a = v(1, a) - v(1, 1)^a$$

Hence v is a symmetric 2-cocycle if and only if $v(1, a) = v(1, 1)^a$, and there are two cocycles:

v	$v(1, 1)$	$v(1, a)$
v_1	0	0
v_2	g	h

Therefore $Z_S^2(S, \mathcal{G}) = B_S^2(S, \mathcal{G})$ and $H_S^2(S, \mathcal{G}) = 0$: every commutative group co-extension of S by $\mathcal{G}$ splits, including Example 1.   □

As this example shows, computing $H_S^2(S, \mathcal{G})$ from its definition is not efficient: just imagine the above if the groups G_1, G_a, ... were larger. We will return to Example 2 in Chap. 6, armed with better methods.

3.2 Comparison with Beck Cohomology

We now compare the symmetric cohomology of S and its Beck cohomology.

3.2.1 Dimension 1

First, the groups $C_B^1(S, \mathcal{G})$ and $C_S^1(S, \mathcal{G})$ are identical. We saw in Sect. 2.3 that $u \in C_B^1(S, \mathcal{G})$ is a Beck 1-cocycle if and only if $u(ab) = u(a)^b + u(b)^a$ for all $a, b \in S$. Hence $Z_S^1(S, \mathcal{G}) = Z_B^1(S, \mathcal{G})$. Thus, $H_S^1(S, \mathcal{G}) = H_B^1(S, \mathcal{G})$.

3.2.2 Dimension 2

Comparing Theorems 2.3.4 and 3.1.1 shows that there is a one-to-one correspondence between $H_S^2(S, \mathcal{G})$ and $H_B^2(S, \mathcal{G})$. This correctly suggests that the two groups are isomorphic:

Theorem 3.2.1 *There is an isomorphism $H_S^2(S, \mathcal{G}) \cong H_B^2(S, \mathcal{G})$ which is natural in $\mathcal{G}$.*

As in [28] and [31] we conduct a rather technical analysis of 2-cocycles, which anticipates the proofs of Theorems 3.2.2 and 3.2.3 below, and shows that the two cohomologies have essentially the same 2-cocycles:

Proposition 3.2.1 *There is an isomorphism $Z_S^2(S, \mathcal{G}) \cong Z_B^2(S, \mathcal{G})$ which is natural in $\mathcal{G}$.*

Proof By (2.30), a Beck 2-cocycle $z \in Z_B^2(S, \mathcal{G})$ is a Beck 2-cochain z such that

$$z(A_1 A_2 \cdots A_m) = z[a_1, a_2, \ldots, a_m] + \sum_{1 \le i \le m} z(A_i)^{a_i^*} \tag{BZ}$$

for all $[A_1, A_2, \ldots, A_m] \in T_2$, where $a_i = \pi_1 A_i$.

A *trimming homomorphism* $\sigma : Z_B^2(S, \mathcal{G}) \longrightarrow C_S^2(S, \mathcal{G})$ is defined by

$$(\sigma z)(a, b) = z[a, b]$$

for all $z \in Z_B^2(S, \mathcal{G})$ and all $a, b \in S$; σz is symmetric since $[b, a] = [a, b]$. □

Lemma 3.2.1 *If z is a Beck 2-cocycle, then $z[a] = 0$ for all $a \in S$ and*

$$z[a_1, a_2, \ldots, a_n] = \sum_{1 < i \le n} z[\overleftarrow{a_i}, a_i]^{\overrightarrow{a_i}} \tag{BZa}$$

for all $a_1, a_2, \ldots, a_n \in S$, where $n \ge 2$,

$$\overleftarrow{a_i} = \begin{cases} 1 \in S^1 & \text{if } i = 1, \\ a_1 \cdots a_{i-1} & \text{if } i \ge 2, \end{cases} \qquad \overrightarrow{a_i} = \begin{cases} a_{i+1} \cdots a_n & \text{if } i \le n - 1, \\ 1 \in S^1 & \text{if } i = n \end{cases}$$

(so that $\overleftarrow{a_i} \, \overrightarrow{a_i} = a_i^*$). Hence σ is injective.

Proof With $m = 1$, $a \in S$, and $A_1 = [a] \in T_1$, (BZ) reads: $z[a] = z[a] + z[a]$, so that $z[a] = 0$.

Now let $n \ge 2$, $a_1, a_2, \ldots, a_n \in S$, and $i > 1$. With $m = 2$, $A_1 = [a_1, a_2, \ldots, a_{i-1}]$, and $A_2 = [a_i]$, so that $\pi A_1 = \overleftarrow{a_i}$, (BZ) yields:

$$z[a_1, a_2, \ldots, a_i] = z[\overleftarrow{a_i}, a_i] + z[a_1, a_2, \ldots, a_{i-1}]^{a_i} + z[a_i]^{\overleftarrow{a_i}}$$

and

$$z[a_1, a_2, \ldots, a_i] = z[a_1, a_2, \ldots, a_{i-1}]^{a_i} + z[\overleftarrow{a_i}, a_i] \tag{BZb}$$

Property (BZa) is now proved by induction on n, using (BZb). If $n = 2$, then (BZa) reduces to (BZb). If (BZa) holds for n, then (BZb) and the induction hypothesis yield

$$
\begin{aligned}
z[a_1, a_2, \ldots, a_{n+1}] &= z[a_1, a_2, \ldots, a_n]^{a_{b+1}} + z[\overleftarrow{a}_{n+1}, a_{n+1}] \\
&= \sum_{1 < i \le n} z[\overleftarrow{a}_i, a_i]^{\overrightarrow{a}_i \, a_{n+1}} + z[\overleftarrow{a}_{n+1}, a_{n+1}] \\
&= \sum_{1 < i \le n+1} z[\overleftarrow{a}_i, a_i]^{\overrightarrow{a}_i{}'}
\end{aligned}
$$

where $\overrightarrow{a}_i{}' = \overrightarrow{a}_i \, a_{n+1} = a_{i+1} \cdots a_n \, a_{n+1}$ if $i < n + 1$, $\overrightarrow{a}_i{}' = 1 \in S^1$ if $i = n + 1$. Thus (BZa) holds for $n + 1$.

It follows from (BZa) that z is completely determined by its values $z[a, b]$ with $a, b \in S$; thus σ is injective.   □

Lemma 3.2.2 *If z is a Beck 2-cocycle, then σz is a symmetric 2-cocycle.*

Proof With $m = 2$, $A_1 = [a, b]$, and $A_2 = [c]$, (BZ) reads

$$
z[a, b, c] = z[ab, c] + z[a, b]^c + z[c]^{ab} = z[ab, c] + z[a, b]^c.
$$

With $m = 2$, $A_1 = [a]$, and $A_2 = [b, c]$, (BZ) reads

$$
z[a, b, c] = z[a, bc] + z[a]^c + z[b, c]^a = z[a, bc] + z[b, c]^a.
$$

Hence σz is a symmetric 2-cocycle.   □

We can now view σ as a homomorphism of $Z_B^2(S, \mathcal{G})$ into $Z_S^2(S, \mathcal{G})$.

Lemma 3.2.3 *The homomorphism $\sigma : Z_B^2(S, \mathcal{G}) \longrightarrow Z_S^2(S, \mathcal{G})$ is surjective.*

Proof Given a symmetric 2-cocycle s, (BZa) suggests (actually, demands) that we define $v(a) = 0$ for all $a \in S$ and

$$
v(a_1, a_2, \ldots, a_n) = \sum_{1 < i \le n} s[\overleftarrow{a}_i, a_i]^{\overrightarrow{a}_i} \tag{V}
$$

for all $n \ge 2$ and all $a_1, a_2, \ldots, a_n \in S$, where $\overleftarrow{a}_i, \overrightarrow{a}_i$ are as in Lemma 3.2.1. Then $v(a, b) = s(a, b)$ and

$$
v(a_1, a_2, \ldots, a_n, a_{n+1}) = \sum_{1 < i \le n} s(\overleftarrow{a}_i, a_i)^{\overrightarrow{a}_i \, a_{n+1}} + s(\overleftarrow{a}_n, a_{n+1})
$$

so that

$$v(a_1, a_2, \ldots, a_n, a_{n+1}) = v(a_1, a_2, \ldots, a_n)^{a_{n+1}} + v(\overleftarrow{a}_n, a_{n+1}) \qquad \text{(W)}$$

We show by induction on n that v has property

$$v(a_{\tau 1}, \ldots, a_{\tau n}) = v(a_1, \ldots, a_n) \qquad \text{(P)}$$

for every permutation τ of $1, \ldots, n$. Since τ is a product of transpositions $(j\ j+1)$, it suffices to prove (P) when $\tau = (j\ j+1)$. If $j \le n-2$, then (P) follows from the induction hypothesis. If $j = n-1$, then, with $\overleftarrow{a}_{n-1} = b_n$, $a_{n-1} = c$, and $a_n = d$, (W), applied twice, yields

$$v(a_1, \ldots, a_{n-2}, c, d) = v(a_1, \ldots, a_{n-2})^{cd} + v(b_n, c)^d + v(b_n c, d)$$

$$v(a_1, \ldots, a_{n-2}, d, c) = v(a_1, \ldots, a_{n-2})^{cd} + v(b_n, d)^c + v(b_n d, c)$$

and (P) follows from (Z2): $s(b, c)^a - s(ab, c) + s(a, bc) - s(a, b)^c = 0$.

By (P), a Beck 2-cochain, also denoted by v, is well defined by:

$$v[a_1, \ldots, a_n] = v(a_1, \ldots, a_n).$$

Property (V) now coincides with (BZa).

Let $A = [a_1, \ldots, a_m]$, $a = a_1 \cdots a_m$, $B = [b_1, \ldots, b_n]$, and $b = b_1 \cdots b_n$. By (V),

$$v[AB] = \sum_{1 < i \le m} s(\overleftarrow{a}_i, a_i)^{\overrightarrow{a}_i b} + s(a, b_1)^{\overrightarrow{b}_1} + \sum_{1 < j \le n} s(a\overrightarrow{b}_j, b_j)^{\overrightarrow{b}_j}$$

so that

$$v[AB] = v[A]^b + \sum_{1 \le j \le k} s(a\overleftarrow{b}_j, b_j)^{\overrightarrow{b}_j} \qquad \text{(BZx)}$$

We can now show, by induction on m, that v has property (Z):

$$z(A_1 A_2 \cdots A_m) = z[a_1, a_2, \ldots, a_m] + \sum_{1 \le i \le m} z(A_i)^{a_i^*}$$

for all $[A_1, A_2, \ldots, A_m] \in T_2$, where $a_i = \pi_1 A_i$. If $m = 1$, (Z) follows from $v[a] = 0$. In general, if $A_1 \cdots A_m = A$, $\pi_1 A_i = a_i$, $a = a_1 \cdots a_m$, and $A_{m+1} =$

$B = [b_1, \ldots, b_k]$, then (BZx), (BZ2), and the induction hypothesis yield

$$v[A_1 \cdots A_m\, A_{m+1}] = v[A]^b + \sum_{1 \le j \le k} s\,(a\,\overleftarrow{b}_j,\, b_j)^{\overrightarrow{b}_j}$$

$$= v[a_1, \ldots, a_m]^b + \sum_{1 \le i \le m} z(A_i)^{a_i^* b} + \sum_{1 \le j \le k} s\,(a\,\overleftarrow{b}_j,\, b_j)^{\overrightarrow{b}_j}$$

$$= v[a_1, \ldots, a_m]^b + \sum_{1 \le i \le m} z(A_i)^{a_i^* b} + s\,(a, b_1)^{\overrightarrow{b}_1}$$

$$+ \sum_{1 < j \le k} \left(-\, s\,(a,\, \overleftarrow{b}_j)^{b_j} + s\,(a,\, \overleftarrow{b}_j\, b_j) + s\,(\overleftarrow{b}_j,\, b_j)^a \right)^{\overrightarrow{b}_j},$$

so that

$$v[A_1 \cdots A_m\, A_{m+1}] = v[a_1, \ldots, a_m]^b + \sum_{1 \le i \le m} z(A_i)^{a_i^* b} + s\,(a, b_1)^{\overrightarrow{b}_1}$$

$$- \sum_{1 < j \le k} \left(s\,(a,\, \overleftarrow{b}_j)^{b_j\,\overrightarrow{b}_j} \right) + \sum_{1 < j \le k} \left(s\,(a,\, \overleftarrow{b}_j\, b_j)^{\overrightarrow{b}_j} \right)$$

$$+ \sum_{1 < j \le k} \left(s\,(\overleftarrow{b}_j,\, b_j)^{a\,\overrightarrow{b}_j} \right).$$

Since $\overleftarrow{b}_j\, b_j = \overleftarrow{b}_{j+1}$ and $b_j\, \overrightarrow{b}_j = \overrightarrow{b}_{j-1}$, the sum

$$s\,(a, b_1)^{\overrightarrow{b}_1} - \sum_{1 < j \le k} \left(s\,(a,\, \overleftarrow{b}_j)^{b_j\,\overrightarrow{b}_j} \right) + \sum_{1 < j \le k} \left(s\,(a,\, \overleftarrow{b}_j\, b_j)^{\overrightarrow{b}_j} \right)$$

cancels down to $s\,(a,\, \overleftarrow{b}_k\, b_k) = s\,(a, b)$. Also

$$\sum_{1 < j \le k} \left(s\,(\overleftarrow{b}_j,\, b_j)^{a\,\overrightarrow{b}_j} \right) = v(B)^a$$

by (V). Since $B = A_{m+1}$ and $b = \pi_1 A_{m+1} = a_{m+1}$, we obtain

$$v[A_1 \cdots A_m\, A_{m+1}] = v[a_1, \ldots, a_m]^b + \sum_{1 \le i \le m} v\,(A_i)^{a_i^* b}$$

$$+ v\,(a, b) + v(B)^a$$

$$= v[a_1, \ldots, a_{m+1}] + \sum_{1 \le i \le m+1} z(A_i)^{a_i'^*}$$

by (W), where $a_i'^* = a_1 \cdots a_{i-1}\, a_{i+1} \cdots a_{m+1}$. This proves (Z).   □

It follows from Lemmas 3.2.1, 3.2.2, and 3.2.3 that $\sigma : Z_B^2(S, \mathcal{G}) \longrightarrow Z_S^2(S, \mathcal{G})$ is an isomorphism. It is readily seen that σ is natural in $\mathcal{G}$. $\qquad\square$

To complete the proof of Theorem 3.2.1 we show that σ preserves coboundaries. Let $u \in C_S^1(S, \mathcal{G}) = C_B^1(S, \mathcal{G})$. By (2.27), the coboundary $\delta^B u$ of u as a Beck 1-cochain is

$$(\delta^B u)[a_1, a_2, \ldots, a_n] = u(a_1 a_2 \cdots a_n) - \sum_{1 \le i \le n} u(a_i)^{a_i^*}$$

for all $a_1, a_2, \ldots, a_n \in S$. Hence

$$(\sigma \delta^B u)(a, b) = u(ab) - u(a)^b - u(b)^a = -(\delta^S u)(a, b)$$

for all $a, b \in S$, where $\delta^S u$ is the coboundary of u as a symmetric 1-cochain. Therefore $\sigma B_B^2(S, \mathcal{G}) = B_S^2(S, \mathcal{G})$, and σ induces an isomorphism

$$H_B^2(S, \mathcal{G}) = Z_B^2(S, \mathcal{G})/B_B^2(S, \mathcal{G}) \cong Z_S^2(S, \mathcal{G})/B_S^2(S, \mathcal{G}) = H_S^2(S, \mathcal{G})$$

which is, like σ, natural in $\mathcal{G}$. $\qquad\square$

3.2.3 Dimensions 3 and 4

How to define symmetric n-cochains when $n > 2$ is a major problem of symmetric cohomology. It is easy to find symmetry properties of the coboundaries of symmetric n-cochains. It is not so easy to find conditions that ensure $H_S^n(S, \mathcal{G}) \cong H_B^n(S, \mathcal{G})$.

In [18] this difficulty was swept under the rug. It was overcome in [28] for $n = 3$ at the cost of a large calculation that proved:

Theorem 3.2.2 *There is an isomorphism $H_S^3(S, \mathcal{G}) \cong H_B^3(S, \mathcal{G})$ which is natural in $\mathcal{G}$.*

The proof resembles the proof of Theorem 3.2 but is substantially longer and uses three trimming homomorphisms. Also it seems that $Z_B^3(S, \mathcal{G})$ and $Z_S^3(S, \mathcal{G})$ are not isomorphic. Theorem 3.2.2 implies that (S3) is an appropriate definition of symmetric 3-cochains.

That (S4) is an appropriate definition of symmetric 4-cochains was proved in a similar fashion by the monstrous calculation in [32]:

Theorem 3.2.3 *There is an isomorphism $H_S^4(S, \mathcal{G}) \cong H_B^4(S, \mathcal{G})$ which is natural in $\mathcal{G}$.*

This result was announced in [31], but the original proof had a few gaps; a horrible but presumably correct proof did not appear until [32]. It resembles the proof of Theorems 3.2.1 and 3.2.2 but is much longer and uses no fewer than nine trimming homomorphisms.

As of this writing, appropriate symmetry conditions have not been established in dimensions $n \geq 5$. Kurdiani and Pirashvili [38] write that they have solved the author's 'cocycle problem' but provide no symmetry conditions.

3.3 Main Properties

The properties of Beck cohomology in Sect. 2.2, combined with Theorems 3.2.1, 3.2.2, and 3.2.3, yield additional properties of H_S^n, as it now exists in dimensions $n \leq 4$.

Theorem 3.3.1 *If S is a free commutative semigroup, then $H_S^n(S, \mathcal{G}) = 0$ for all $\mathcal{G}$, where $n = 2, 3, 4$.*

This follows from Theorem 2.3.2. If $n = 2$, then Theorem 3.1.1 offers a more direct proof: if S is free, then every commutative group coextension of S must split.

Theorem 3.3.2 *Every short exact sequence $\mathcal{S}: 0 \longrightarrow \mathcal{G} \longrightarrow \mathcal{G}' \longrightarrow \mathcal{G}'' \longrightarrow 0$ of abelian group valued functors on S induces an exact sequence*

$$0 \longrightarrow H_S^1(S, \mathcal{G}) \longrightarrow H_S^1(S, \mathcal{G}') \longrightarrow H_S^1(S, \mathcal{G}'')$$

$$\longrightarrow H_S^2(S, \mathcal{G}) \longrightarrow H_S^2(S, \mathcal{G}') \longrightarrow H_S^2(S, \mathcal{G}'')$$

$$\longrightarrow H_S^3(S, \mathcal{G}) \longrightarrow H_S^3(S, \mathcal{G}') \longrightarrow H_S^3(S, \mathcal{G}'')$$

$$\longrightarrow H_S^4(S, \mathcal{G}) \longrightarrow H_S^4(S, \mathcal{G}') \longrightarrow H_S^4(S, \mathcal{G}'')$$

which is natural in $\mathcal{S}$.

This follows from Theorem 2.3.3. A direct proof will be found in Chap. 7.

These properties, together with Theorem 3.1.1, a more natural relationship with group coextensions, and very manageable cochains, give the symmetric cohomology a strong claim to be the 'right' cohomology of commutative semigroups; or would give, were it defined in all dimensions.

3.4 Normalization

Since every commutative monoid is a commutative semigroup, the cohomology of commutative semigroups is 'obviously' more general than that of commutative monoids. This has led the author to neglect the latter (except in dimension 2). But this section will show that, in dimensions 2 and 3, the cohomology of commutative monoids is 'obviously' more general than that of commutative semigroups.

This follows from properties of *normalized* cochains: if S has an identity element 1, then an n-cochain $u \in C_S^n(S, \mathcal{G})$ is *normalized* if and only if $u(a_1, a_2, \ldots, a_n) = 0$ whenever $a_i = 1$ for some i.

In particular, if S has an identity element, then a 1-cochain $u \in C^1(S, \mathcal{G})$ is *normalized* if and only if $u(1) = 0$; a symmetric 2-cochain $u \in C_S^2(S, \mathcal{G})$ is *normalized* if and only if $u(1, a) = 0$ for all $a \in S$. (A more general definition, of interest if S is finite, is given in [31], according to which $u \in C_S^2(S, \mathcal{G})$ is normalized if $u(e, a) = 0$ whenever $e^2 = e$ and $ea = a$.)

Under pointwise addition, normalized n-cochains constitute an abelian group $C_N^n(S, \mathcal{G}) \subseteq C^n(S, \mathcal{G})$; if $n = 1, 2, 3, 4$, then normalized symmetric n-cochains constitute an abelian group $C_{NS}^n(S, \mathcal{G}) = C_N^n(S, \mathcal{G}) \cap C_S^n(S, \mathcal{G})$, and normalized symmetric n-cocycles constitute an abelian group $Z_{NS}^n(S, \mathcal{G}) = C_N^n(S, \mathcal{G}) \cap Z_S^n(S, \mathcal{G})$.

3.4.1 Dimension 2

If $s \in Z_S^2(S, \mathcal{G})$, and S has an identity element, then applying (Z2): $s(b, c)^a - s(ab, c) + s(a, bc) - s(a, b)^c = 0$ to $1, 1, a$ yields

$$s(1, 1)^a - s(11, a) + s(1, 1a) - s(1, a)^1 = 0$$

and

$$s(1, a) = s(1, 1)^a \tag{3.4}$$

for all $a \in S$. Hence a symmetric 2-cocycle s is normalized if and only if $s(1, 1) = 0$.

Lemma 3.4.1 *If S is a monoid, then* $Z_S^2(S, \mathcal{G}) = Z_{NS}^2(S, \mathcal{G}) + B_S^2(S, \mathcal{G})$.

Proof Let $s \in Z_S^2(S, \mathcal{G})$. Define a 1-cochain $u \in C^1(S, \mathcal{G})$ by

$$u(a) = s(1, a)$$

for all $a \in S$. Then

$$(s - \delta u)(1, a) = s(1, a) - u(a)^1 + u(1a) - u(1)^a = s(1, a) - s(1, 1)^a = 0$$

for all $a \in S$, by (3.4), and $s - \delta u$ is normalized. □

Normalized 2-coboundaries can be defined either as 2-coboundaries that are normalized, or as the coboundaries of normalized 1-cochains. The two definitions are equivalent:

Lemma 3.4.2 *If S is a monoid, then* $B_S^2(S, \mathcal{G}) \cap C_N^2(S, \mathcal{G}) = \delta\, C_N^1(S, \mathcal{G})$.

Proof If $u \in C^1(S, \mathcal{G})$ is normalized, then

$$(\delta u)(1, a) = u(a)^1 - u(1a) + u(1)^a = 0$$

and δu is normalized. Conversely, if δu is normalized, then

$$u(1) = u(1)^1 - u(11) + u(1)^1 = (\delta u)(1, 1) = 0$$

and u is normalized.   □

We can now define

$$B_{NS}^2(S, \mathcal{G}) = B_S^2(S, \mathcal{G}) \cap C_N^2(S, \mathcal{G}) = \delta C_N^1(S, \mathcal{G}).$$

Now let S be any commutative semigroup, and embed S into a monoid $S' = S \cup \{1\}$ whose multiplication extends that of S. If S is already blessed with an identity element, then the added identity element 1 duplicates the existing identity element. If S is not blessed with an identity element, then $S' = S^1$. In either case, every abelian group valued functor $\mathcal{G}$ on S extends to an abelian group valued functor $\mathcal{G}'$ on S', in which $G_1' = 0$. We use this construction to prove:

Theorem 3.4.1 *For every commutative semigroup S and abelian group valued functor $\mathcal{G}$ on S there is an isomorphism*

$$H_S^2(S, \mathcal{G}) \cong H_S^2(S', \mathcal{G}')$$

which is natural in $\mathcal{G}$. If S has an identity element, then there is for every abelian group valued functor $\mathcal{G}$ on S an isomorphism

$$H_S^2(S, \mathcal{G}) \cong Z_{NS}^2(S, \mathcal{G}) / B_{NS}^2(S, \mathcal{G})$$

which is natural in $\mathcal{G}$.

Proof If S has an identity element, then $H_S^2(S, \mathcal{G}) \cong Z_{NS}^2(S, \mathcal{G}) / B_{NS}^2(S, \mathcal{G})$ follows from $Z_S^2(S, \mathcal{G}) = Z_{NS}^2(S, \mathcal{G}) + B_S^2(S, \mathcal{G})$ in Lemma 3.4.1 and $B_{NS}^2(S, \mathcal{G}) = B_S^2(S, \mathcal{G}) \cap Z_{NS}^2(S, \mathcal{G})$ in Lemma 3.4.2.

Now do not assume that S has an identity element. By the above,

$$H_S^2(S', \mathcal{G}') \cong Z_{NS}^2(S', \mathcal{G}') / B_{NS}^2(S', \mathcal{G}').$$

A normalized 2-cocycle $s' \in Z_{NS}^2(S', \mathcal{G}')$ is completely determined by its values $s'(a, b)$ where $a, b \neq 1$, which constitute a 2-cocycle $s \in Z_S^2(S, \mathcal{G})$. If $s' = \delta u'$ is a coboundary, then so is $s = \delta u$, where u is the restriction of u' to S.

Conversely, every 2-cocycle $s \in Z_S^2(S, \mathcal{G})$ extends to a normalized 2-cocycle s' on S^1, to wit

$$s'(a, b) = \begin{cases} s(a, b) & \text{if } a, b \in S, \\ 0 & \text{if } a = 1 \text{ or if } b = 1. \end{cases}$$

Property (Z2): $s(b, c)^a - s(ab, c) + s(a, bc) - s(a, b)^c = 0$ of s extends to s': this is clear if $a, b, c \in S$; if $a = 1$, then

$$s'(b, c)^1 - s'(1b, c) + s'(1, bc) - s'(1, b)^c = 0;$$

if $b = 1$, then

$$s'(1, c)^a - s'(a1, c) + s'(a, 1c) - s'(a, 1)^c = 0;$$

if $c = 1$, then

$$s'(b, 1)^a - s'(ab, 1) + s'(a, b1) - s'(a, b)^1 = 0.$$

If in the above $s = \delta u$ is a coboundary, then u extends to a normalized 1-cochain u' on S^1 and $s' = \delta u'$ is also a coboundary.

We now have an isomorphism $Z_S^2(S, \mathcal{G}) \longrightarrow Z_{NS}^2(S', \mathcal{G}')$, $s \longmapsto s'$, which is natural in $\mathcal{G}$, sends $B_S^2(S, \mathcal{G})$ onto $B_{NS}^2(S', \mathcal{G}')$, and induces an isomorphism $H_S^2(S, \mathcal{G}) \longrightarrow H_S^2(S', \mathcal{G}')$, which is also natural in $\mathcal{G}$. □

3.4.2 Dimension 3

Similar results hold in dimension 3. If S has an identity element, then a symmetric 3-cochain $t \in C^3(S, \mathcal{G})$ is normalized if and only if $t(1, a, b) = 0$ for all $a, b \in S$, for then properties (S3a) and (S3b) imply $t(a, 1, b) = t(a, b, 1) = 0$ for all $a, b \in S$.

Lemma 3.4.3 *If S is a monoid, then $Z_S^3(S, \mathcal{G}) = Z_{NS}^3(S, \mathcal{G}) + B_S^3(S, \mathcal{G})$.*

Proof Let $t \in Z_S^3(S, \mathcal{G})$ be a symmetric 3-cocycle. Since

$$(\delta t)(1, 1, b, c) = t(1, b, c)^1 - t(11, b, c) + t(1, 1b, c) - t(1, 1, bc) + t(1, 1, b)^c$$

we have

$$t(1, b, c) - t(1, 1, bc) + t(1, 1, b)^c = 0$$

for all $b, c \in S$. (Hence t is normalized if and only if $t(1, 1, a) = 0$ for all $a \in S$.)
Define

$$\begin{cases} u(1, a) = u(a, 1) = t(1, 1, a) & \text{for all } a \in S \\ u(a, b) = 0 & \text{if } a, b \neq 1 \end{cases}$$

for all $a, b \in S$. Then u is a symmetric 2-cochain. Moreover,

$$\begin{aligned} (t - \delta u)(1, b, c) &= t(1, b, c) - u(b, c)^1 + u(1b, c) - u(1, bc) + u(1, b)^c \\ &= t(1, b, c) - t(1, 1, bc) + t(1, 1, b)^c = 0 \end{aligned}$$

for all $b, c \in S$. Hence $t - \delta u$ is normalized.   □

As in dimension 2, normalized 3-coboundaries can be defined either as 3-co-boundaries that are normalized, or as the coboundaries of normalized 2-cochains. The two definitions are equivalent:

Lemma 3.4.4 *If S is a monoid, then $B_S^3(S, \mathcal{G}) \cap C_N^3(S, \mathcal{G}) = \delta\, C_{NS}^2(S, \mathcal{G})$.*

Proof Let $t = \delta u$ for some $u \in C_S^2(S, \mathcal{G})$. If u is normalized, then

$$t(1, b, c) = u(b, c)^1 - u(1b, c) + u(1, bc) - u(1, b)^c = 0$$

for all $b, c \in S$, and t is normalized.

Conversely assume that t is normalized. Then $u(1, bc) = u(1, b)^c$ for all $b, c \in S$; in particular, $u(1, a) = u(1, 1)^a$ for all $a \in S$. If

$$v(a) = u(1, a)$$

for all $a \in S$, then

$$(u - \delta v)(1, a) = u(1, a) - v(a)^1 + v(1a) - v(1)^a = u(1, a) - u(1, 1)^a = 0$$

for all $a \in S$, so that $u - \delta v$ is normalized. Moreover, $t = \delta u = \delta(u - \delta v)$.   □

We can now define

$$B_{NS}^3(S, \mathcal{G}) = B_S^3(S, \mathcal{G}) \cap C_N^3(S, \mathcal{G}) = \delta\, C_{NS}^2(S, \mathcal{G}).$$

As above, every abelian group valued functor $\mathcal{G}$ on S extends to an abelian group valued functor $\mathcal{G}'$ on $S' = S \cup \{1\}$, such that $G_1' = 0$.

Theorem 3.4.2 *For every commutative semigroup S and abelian group valued functor $\mathcal{G}$ on S there is an isomorphism*

$$H_S^3(S, \mathcal{G}) \cong H_S^3(S', \mathcal{G}')$$

which is natural in $\mathcal{G}$. If S has an identity element, then there is for every abelian group valued functor $\mathcal{G}$ on S an isomorphism

$$H_S^3(S, \mathcal{G}) \cong Z_{NS}^3(S, \mathcal{G}) / B_{NS}^3(S, \mathcal{G})$$

which is natural in $\mathcal{G}$.

Proof If S has an identity element, then $H_S^3(S, \mathcal{G}) \cong Z_{NS}^3(S, \mathcal{G}) / B_{NS}^3(S, \mathcal{G})$ follows from $Z_S^3(S, \mathcal{G}) = Z_{NS}^3(S, \mathcal{G}) + B_S^3(S, \mathcal{G})$ in Lemma 3.4.3 and $B_{NS}^3(S, \mathcal{G}) = B_S^3(S, \mathcal{G}) \cap Z_{NS}^3(S, \mathcal{G})$.

Now do not assume that S has an identity element. By the above,

$$H_S^3(S', \mathcal{G}') \cong Z_{NS}^3(S', \mathcal{G}') / B_{NS}^3(S', \mathcal{G}').$$

A normalized 3-cocycle $t' \in Z_{NS}^3(S', \mathcal{G}')$ is completely determined by its values $t'(a, b, c)$ where $a, b, c \neq 1$, which constitute a 3-cocycle $t \in Z_S^3(S, \mathcal{G})$. If $t' = \delta u'$ is a coboundary, then so is $t = \delta u$, where u is the restriction of u' to $S \times S$.

Conversely, every 3-cocycle $t \in Z_S^3(S, \mathcal{G})$ extends to a normalized 3-cocycle t' on S'. Indeed let

$$t'(a, b, c) = \begin{cases} t(a, b, c) & \text{if } a, b, c \in S, \\ 0 & \text{if } a = 1, b = 1, \text{ or } c = 1. \end{cases}$$

Then t' is symmetric, like t. Property (Z3) of t:

$$t(b, c, d)^a - t(ab, c, d) + t(a, bc, d) - t(a, b, cd) + t(a, b, c)^d = 0$$

also extends to t': this is clear if $a, b, c, d \in S$; if $a = 1$, then

$$t'(b, c, d)^1 - t'(1b, c, d) + t'(1, bc, d) - t'(1, b, cd) + t'(1, b, c)^d = 0;$$

if $b = 1$, then

$$t'(1, c, d)^a - t'(a1, c, d) + t'(a, 1c, d) - t'(a, 1, cd) + t'(a, 1, c)^d = 0;$$

the cases $c = 1$ and $d = 1$ then follow from (S3a) and the property (S4a): $u(d, c, b, a) = -u(a, b, c, d)$ of $\delta t'$.

If in the above $t = \delta u$ is a coboundary, then u extends to a normalized 3-cochain u' on S' and $t' = \delta u'$ is also a coboundary.

We now have an isomorphism $Z_S^3(S, \mathcal{G}) \longrightarrow Z_{NS}^3(S', \mathcal{G}')$, $t \longmapsto t'$, which is natural in $\mathcal{G}$, sends $B_S^3(S, \mathcal{G})$ onto $B_{NS}^3(S', \mathcal{G}')$, and induces an isomorphism $H_S^3(S, \mathcal{G}) \longrightarrow H_S^3(S', \mathcal{G}')$, which is also natural in $\mathcal{G}$. $\qquad\square$

The isomorphisms $H_S^2(S, \mathcal{G}) \cong H_S^2(S', \mathcal{G}')$ and $H_S^3(S, \mathcal{G}) \cong H_S^3(S', \mathcal{G}')$ in Theorems 3.4.1 and 3.4.2 show that, in dimensions 2 and 3, the symmetric cohomology of commutative semigroups in general is completely determined by the symmetric cohomology of those commutative monoids that, like S', have a trivial group of units. Therefore the symmetric cohomology of commutative monoids is 'obviously' more general than that of commutative semigroups. But then Theorems 3.4.1 and 3.4.2 cannot even be stated unless there is a cohomology that does not assume an identity element.

In dimensions $n \leq 3$ we now may, without loss of generality, assume that S has an identity element. It seems likely that Theorems 3.4.1 and 3.4.2 extend to dimension 4 and to the Beck cohomology, but at this time the author does not have a proof.

Chapter 4
Calvo-Cegarra Cohomology

Calvo-Cegarra cohomology is the cohomology of simplicial objects constructed from commutative monoids. It has good properties and agrees with the Beck cohomology and the symmetric cohomology in dimensions 1 and 2, so that it classifies finite commutative monoids, but its larger groups in dimensions $n \geq 3$ hold more information. This gives the Calvo-Cegarra cohomology a strong claim to be the 'better' cohomology of commutative monoids.

4.1 Small Categories

A *small* category is a category whose class of objects and class of morphisms are sets. A cohomology was defined by Roos [49] for every small category $\mathcal{C}$.

Abelian group valued functors on $\mathcal{C}$ are the objects of an abelian category $\mathcal{A}$. The limit functor $\mathbb{L}$ that sends an abelian group valued functor $\mathcal{G}$ to its limit $\mathbb{L}(\mathcal{G}) = \lim \mathcal{G}$ is itself an abelian group valued functor on $\mathcal{A}$. Since $\mathcal{A}$ has enough injective objects, the left exact functor $\mathbb{L}$ has right derived functors $H^n(\mathcal{C}, -)$, which provide the *cohomology groups* $H^n(\mathcal{C}, \mathcal{G})$ of $\mathcal{C}$ with coefficients in any abelian group valued functor $\mathcal{G}$ on $\mathcal{C}$.

A long exact sequence for $H^n(\mathcal{C}, -)$ follows from the similar property of right derived functors:

Theorem 4.1.1 *If $\mathcal{C}$ is a small category, then for each short exact sequence*

$$\mathcal{S}: 0 \longrightarrow \mathcal{G} \longrightarrow \mathcal{G}' \longrightarrow \mathcal{G}'' \longrightarrow 0$$

© The Author(s), under exclusive license to Springer Nature Switzerland AG 2022   45
P. A. Grillet, *The Cohomology of Commutative Semigroups*, Lecture Notes
in Mathematics 2307, https://doi.org/10.1007/978-3-031-08212-2_4

of abelian group valued functors on $\mathcal{C}$ *there is an exact sequence*

$$0 \longrightarrow \lim \mathcal{G} \longrightarrow \lim \mathcal{G}' \longrightarrow \lim \mathcal{G}'' \longrightarrow H^1(\mathcal{C}, \mathcal{G}) \longrightarrow H^1(\mathcal{C}, \mathcal{G}')$$
$$\cdots$$
$$\longrightarrow H^n(\mathcal{C}, \mathcal{G}') \longrightarrow H^n(\mathcal{C}, \mathcal{G}'') \longrightarrow H^{n+1}(\mathcal{C}, \mathcal{G}) \longrightarrow H^{n+1}(\mathcal{C}, \mathcal{G}') \cdots$$

which is natural in $\mathcal{S}$.

4.2 Cohomology of Simplicial Sets

This cohomology was developed by Gabriel and Zisman [17].

4.2.1 Definition

As in Sect. 9.3, a simplicial set X consists of sets $X_0, X_1, \ldots, X_n, \ldots$, together with face maps $d_i \colon X_n \longrightarrow X_{n-1}$ $(i = 0, 1, \ldots, n)$ and degeneracy maps $s_i \colon X_{n-1} \longrightarrow X_n$ $(i = 0, 1, \ldots, n-1)$, that satisfy all simplicial identities:

$$
\begin{aligned}
d_i \circ d_j &= d_{j-1} \circ d_i && \text{if } i < j \\
d_i \circ s_j &= s_{j-1} \circ d_i && \text{if } i < j \\
d_i \circ s_j &= 1_{X_{n-1}} && \text{if } i = j \text{ or if } i = j+1 \\
d_i \circ s_j &= s_j \circ d_{i-1} && \text{if } i > j+1 \\
s_i \circ s_j &= s_{j+1} \circ s_i && \text{if } i \le j.
\end{aligned}
$$

A simplicial set X can be viewed as a contravariant functor on the simplicial category Δ whose objects are all sets $[n] = \{0, 1 \ldots, n\}$ and whose morphisms $[m] \longrightarrow [n]$ are all monotone mappings of $[m]$ into $[n]$; Δ has face and degeneracy maps δ_i, σ_i that satisfy the opposite simplicial identities: $\delta_j \circ \delta_i = \delta_i \circ \delta_{j-1}$ if $i < j$, and so forth (see Sect. 9.3).

In what follows we assume that the sets X_n are pairwise disjoint, so that n is determined by x when $x \in X_n$. An *n-simplex* of X is an element of X_n; a *simplex* of X is an element of X_n for some n.

The simplexes of a simplicial set X are the objects of a category Δ_X (denoted by Δ/X in [10]), the *category of simplexes* of X, which is somewhat similar to Δ. A morphism of Δ_X from $y \in X_m$ to $x \in X_n$ is a pair (φ, x) in which $\varphi \colon [m] \longrightarrow [n]$ is a morphism of Δ such that $y = X(\varphi)(x)$. For example, $(\delta_i, x) \colon d_i x \longrightarrow x$, where $x \in X_{n+1}$ and $i \le n+1$, and $(\sigma_i, x) \colon s_i x \longrightarrow x$, where $x \in X_n$ and

$i \leq n$, are morphisms of Δ_X, which satisfy all the opposite simplicial identities, $(\delta_j, x) \circ (\delta_i, d_j x) = (\delta_i, x) \circ (\delta_{j-1}, d_i x)$ if $i < j$, and so forth

The category Δ_X inherits the universal property of Δ: if a category C contains objects C_x, one for each $x \in \bigcup_{n \geq 0} X_n$, and morphisms $\delta_{i,x}: C_{d_i x} \longrightarrow C_x$, $\sigma_{i,x}: C_{s_i x} \longrightarrow C_x$ that satisfy all the opposite simplicial identities $\delta_{j,x} \circ \delta_{i,d_j x} = \delta_{i,x} \circ \delta_{j-1,d_i x}$ if $i < j$, and so forth, then there exists a unique functor $F: \Delta_X \longrightarrow C$ such that $Fx = C_x$, $F(\delta_i, x) = \delta_{i,x}$, and $F(\sigma_i, x) = \sigma_{i,x}$, for all $0 \leq i \leq n$ and all $x \in X_n$.

In particular, an abelian group valued functor $\mathcal{G}$ on Δ_X assigns an abelian group G_x to each simplex $x \in \bigcup_{n \geq 0} X_n$ and is completely determined by homomorphisms

$$\delta_{i,x} = \mathcal{G}(\delta_i, x): G_{d_i x} \longrightarrow G_x \text{ and } \sigma_{i,x} = \mathcal{G}(\sigma_i, x): G_{s_i x} \longrightarrow G_x$$

that satisfy all opposite simplicial identities.

The *Gabriel-Zisman cohomology* of a simplicial set X is that of the small category Δ_X; coefficients are provided by abelian group valued functors on Δ_X.

4.2.2 Cochains

The Gabriel-Zisman cohomology groups $H^n(X, \mathcal{G}) = H^n(\Delta_X, \mathcal{G})$ are also the homology groups of a cochain complex [17].

Let $\mathcal{G}$ be an abelian group valued functor on Δ_X. An *n-cochain* u on X with values in $\mathcal{G}$ assigns $u(x) \in G_x$ to each n-simplex $x \in X_n$. Under pointwise addition these n-cochains constitute an abelian group $C^n(X, \mathcal{G})$.

The cochain groups $C^n(X, \mathcal{G})$ come with homomorphisms

$$C^{n-1}(X, \mathcal{G}) \xrightarrow{d_{n,i}} C^n(X, \mathcal{G}) \xleftarrow{s_{n,i}} C^{n+1}(X, \mathcal{G}) \qquad (4.1)$$

for each $0 \leq i \leq n$, defined for all $x \in X_n$ by

$$
\begin{aligned}
(d_{n,i}u)(x) &= \delta_{i,x}(u(d_i x)) = \mathcal{G}(\delta_i, x)(u(d_{n,i}x)), \\
(s_{n,i}u)(x) &= \sigma_{i,x}(u(s_i x)) = \mathcal{G}(\sigma_i, x)(u(s_{n,i}x)).
\end{aligned}
\qquad (4.2)
$$

This arranges the cochain groups $C^n(X, \mathcal{G})$ into a cosimplicial abelian group. As in Sect. 9.3, the *coboundary* of an n-cochain $u \in C^n(X, \mathcal{G})$ is

$$\delta_n u = \sum_{0 \leq i \leq n} (-1)^i d_{n,i} u \in C^{n+1}(X, \mathcal{G}); \qquad (4.3)$$

the simplicial identities imply that $\delta_{n+1} \circ \delta_n = 0$.

An n-cochain u on X is *normalized* if $s_{n,i}u = 0$ for all $i = 0, 1, \ldots, n$. Under pointwise addition, normalized n-cochains on X constitute a subgroup $C^n_N(X, \mathcal{G})$ of

$C^n(X, \mathcal{G})$. If u is normalized, then it follows from the simplicial identities that $d_{n,i}u$ is normalized; hence $\delta_n u$ is normalized, and (4.3) yields a complex

$$0 \longrightarrow C_N^0(X, \mathcal{G}) \longrightarrow \cdots \longrightarrow C_N^n(X, \mathcal{G}) \xrightarrow{\delta_n} C_N^{n+1}(X, \mathcal{G}) \cdots \qquad (4.4)$$

For the sake of completeness, $\delta_{-1} = 0 : 0 \longrightarrow C_N^0(X, \mathcal{G})$.

Theorem 4.2.1 ([17]) *For each $n \geq 0$ there is an isomorphism*

$$H^n(X, \mathcal{G}) = \operatorname{Ker} \delta_n / \operatorname{Im} \delta_{n-1}$$

which is natural in $\mathcal{G}$.

4.2.3 The Classifying Simplicial Set

The *classifying simplicial set* $V = V(S)$ of a monoid S (denoted by $\overline{W} S$ in [10]) is constructed as follows. If $n > 0$, then V_n is the cartesian product

$$V_n = S \times S \times \cdots \times S = S^{(n)}.$$

If $n = 0$, then $V_n = V_0$ is the empty cartesian product, with just one element (). We regard the sets V_n as pairwise disjoint.

The face and degeneracy maps d_i' and s_i' of $V(S)$ are:

$$d_0'(a) = d_1'(a) = (), \quad s_0'() = (1) \qquad\qquad\qquad (4.5)$$

$$d_0'(a_1, \ldots, a_n) = (a_2, \ldots, a_n) \qquad\qquad\qquad \text{if } n > 1$$

$$d_i'(a_1, \ldots, a_n) = (a_1, \ldots, a_{i-1}, a_i a_{i+1}, a_{i+2}, \ldots, a_n) \qquad \text{if } 0 < i < n$$

$$d_n'(a_1, \ldots, a_n) = (a_1, \ldots, a_{n-1}) \qquad\qquad\qquad \text{if } n > 1$$

$$s_i'(a_1, \ldots, a_n) = (a_1, \ldots, a_{i-1}, 1, a_i, \ldots a_n)$$

for all $a \in S$ and $(a_1, \ldots, a_n) \in V_n$. Thus, d_0' removes the first component a_1 of $(a_1, \ldots, a_n)$, just as d_n' removes its last component, whereas the product of all components remains unchanged by s_i' and by all other d_i'.

More generally, for every simplicial monoid M, Eilenberg and MacLane [16] constructed a classifying simplicial set (denoted by WM in [16], by $\overline{W}M$ in [10]). Section 9.3 gives this construction. 'Our' $V(S)$ is the particular case where M is constant at S: $M_n = S$ and $d_i = s_i = 1_S$ for all n and i.

There is a canonical functor of $\Delta_{V(S)}$ to the Leech category $\mathcal{D}(S)$ (in Sect. 9.1) that sends $(a_1, \ldots, a_n) \in V_n$ to $a_1 \cdots a_n \in S$, sends $(d_i', (a_1, \ldots, a_n))$ to

$$
\begin{cases}
(a_1, a_2 \cdots a_n, 1): a_2 \cdots a_n \longrightarrow a_1 \cdots a_n & \text{if } i = 0, \\
(1, a_1 \cdots a_n, 1) = 1_{a_1 \cdots a_n} & \text{if } 0 < i < n, \\
(1, a_1 \cdots a_{n-1}, a_n): a_1 \cdots a_{n-1} \longrightarrow a_1 \cdots a_n & \text{if } i = n,
\end{cases}
$$

and sends $(s_i', (a_1, \ldots, a_n))$ to $(1, a_1 \cdots a_n, 1)$ [10]. Hence an abelian group valued functor $\mathcal{G}$ on the Leech category $\mathcal{D}(S)$ induces an abelian group valued functor $\mathcal{G}' = \mathcal{G}$ on $\Delta_{V(S)}$. Theorem 4.2.1 then implies that $H^n(V(S), \mathcal{G}')$ is isomorphic to the Leech cohomology group $H_L^n(S, \mathcal{G})$.

4.3 Cohomology of Commutative Semigroups

This section defines the Calvo-Cegarra cohomology of commutative semigroups, following Calvo-Cervera and Cegarra [10].

4.3.1 The Double Classifying Simplicial Set

If S is a commutative monoid, then every $V_n = S^n$ is a commutative monoid under componentwise multiplication, commutativity makes all maps d_i', s_i' monoid homomorphisms, and the classifying simplicial set $V(S)$ is in fact a simplicial commutative monoid. Therefore $V(S)$ has a classifying simplicial set $W = W(S)$ (denoted by $\overline{W}^2 S$ in [10]). As in [10], the general construction in [16] and Sect. 9.3 yields $W_0 = \{()\}$; $W_1 = S^{(0)} = \{()\}$; omitting the trivial term $S^{(0)} = \{()\}$,

$$
W_{n+1} = S^{(n)} \times S^{(n-1)} \times \cdots \times S \quad \text{if } n > 0;
$$

and face and degeneracy mappings d_i'', s_i'' obtained by applying (9.39) to d_i', s_i' in (4.5):

$$
\begin{aligned}
d_0''(x_1) &= d_1''(x_1) = (), \quad s_0''() = (1) && (4.6) \\
d_0''(x_n, \ldots, x_1) &= (x_{n-1}, \ldots, x_1) && \text{if } n > 0 \\
d_{i+1}''(x_n, \ldots, x_1) &= (d_i' x_n, \ldots, d_1' x_{n-i+1}, \\
& \qquad (d_0' x_{n-i})(x_{n-i-1}), x_{n-i-2}, \ldots, x_1) && \text{if } i < n \\
d_{n+1}''(x_n, \ldots, x_1) &= (d_n' x_n, \ldots, d_1' x_1) && \text{if } n > 0 \\
s_0''(x_n, \ldots, x_1) &= (1, x_n, \ldots x_1)
\end{aligned}
$$

$$s_{i+1}''(x_n, \dots, x_1) = (s_i' x_n, \dots, s_0' x_{n-i},$$
$$1, x_{n-i-1}, \dots, x_1) \qquad \text{if } i < n$$
$$s_{n+1}''(x_n, \dots, x_1) = (s_n' x_n, \dots, s_1' x_1, 1)$$

for all $x_1 \in S$, $x_2 \in S^{(2)}, \dots, x_n \in S^{(n)}$.

For example,

$$d_0''(a) = d_1''(a) = d_2''(a) = ()$$
$$s_0''(a) = ((1,1),a), \quad s_1''(a) = ((1,a),1), \quad s_2''(a) = ((a,1),1)$$

whereas

$$d_0''((c,b),a) = (a), \quad d_1''((c,b),a) = (ba),$$
$$d_2''((c,b),a) = (cb), \quad d_3''((c,b),a) = (c),$$
$$s_0''((c,b),a) = ((1,1,1),(c,b),a),$$
$$s_1''((c,b),a) = ((1,c,b),(1,1),a),$$
$$s_2''((c,b),a) = ((c,1,b),(1,a),1),$$
$$s_3''((c,b),a) = ((c,b,1),(a,1),1).$$

In general, d_0'' removes the first component x_n of $x = (x_n, \dots, x_1)$, just like d_0' removes the first component y_n of x_n, whereas d_{n+1}'' removes the last components $z_n, \dots, z_1$ of all components $x_n, \dots, x_1$ of x.

There is a canonical projection $\mathbb{P}$ of Δ_W to the Leech category $\mathcal{H}(S)$. For each $n > 0$ let $\pi_n : S^{(n)} \longrightarrow S$ be the product homomorphism

$$\pi_n : (a_1, a_2, \dots, a_n) \longmapsto a_1 a_2 \cdots a_n ;$$

let $\pi_0() = 1$. If $x = (x_n, \dots, x_1) \in W_{n+1}$, then

$$\mathbb{P}(x) = (\pi_n x_n) \cdots (\pi_1 x_1).$$

The values of $\mathbb{P}$ on morphisms are determined by its values at

$$(\delta_i, x) : d_i'' x \longrightarrow x \quad \text{and} \quad (\sigma_i, x) : s_i'' x \longrightarrow x .$$

For all $x = (x_n, \dots, x_1) \in W_{n+1}$:

$$\mathbb{P}(\delta_0, x) = (\mathbb{P}(d_0'' x), \pi_n x_n),$$
$$\mathbb{P}(\delta_{i+1}, x) = (\mathbb{P}(d_{i+1}'' x), y_{n-i}) \qquad \text{if } i < n,$$

$$\mathbb{P}(\delta_{n+1}, x) = (\mathbb{P}(d''_{n+1}x), z_n \cdots z_1),$$

$$\mathbb{P}(\sigma_i, x) = 1_{\mathbb{P}(x)},$$

where y_i is the first component of x_i and z_i is its last component.

Every abelian group valued functor $\mathcal{G}$ on $\mathcal{H}(S)$ now lifts to an abelian group valued functor $\mathcal{G} \circ \mathbb{P}$ on Δ_W.

If S is a commutative monoid and $\mathcal{G}$ is an abelian group valued functor on S, then the *n-th Calvo-Cegarra cohomology group* of S with coefficients in $\mathcal{G}$ is defined as

$$H^n_C(S, \mathcal{G}) = H^{n+1}(W(S), \mathcal{G} \circ \mathbb{P}).$$

The dimension shift that defines $H^n_C(S)$ as $H^{n+1}(W)$ avoids dimension mismatches with previous cohomologies.

4.3.2 Cochains

By Theorem 4.2.1, $H^{n+1}(W, \mathcal{G} \circ \mathbb{P})$ is the $(n+1)$-th homology group of the cochain complex (4.4). To match the dimension shift above, let

$$C^n(W, \mathcal{G}) = C^{n+1}(W, \mathcal{G} \circ \mathbb{P}).$$

Thus an *n*-cochain $u \in C^n(W, \mathcal{G})$ assigns $u(x) \in G_{\mathbb{P}(x)}$ to each element $x = (x_n, \ldots, x_1)$ of W_{n+1}. In particular, a 1-cochain u assigns $u(a) \in G_a$ to each $a \in W_2 = S$; a 2-cochain u assigns $u((c, b), a) \in G_{\mathbb{P}((c,b),a)} = G_{abc}$ to each $((c, b), a) \in W_3 = S^{(2)} \times S$.

Equation (4.2) yields homomorphisms

$$C^{n-1}(W, \mathcal{G}) \xrightarrow{\delta^i} C^n(W, \mathcal{G}) \xleftarrow{\sigma^i} C^{n+1}(W, \mathcal{G})$$

defined (after the dimension shift) by

$$(\delta^0 u)(x) = \mathcal{G}(\mathbb{P}(\delta_0), x)(u(d''_0 x)) = u(d''_0 x)^{\pi_n x_n},$$

$$(\delta^i u)(x) = \mathcal{G}(\mathbb{P}(\delta_{i+1}), x)(u(d''_{i+1} x)) = u(d''_{i+1} x)^{y_{n-1}} \qquad \text{if } i < n$$

$$(\delta^n u)(x) = \mathcal{G}(\mathbb{P}(\delta_{n+1}), x)(u(d''_{n+1} x)) = u(d''_{n+1} x)^{z_n \cdots z_1}$$

$$(\sigma^i u)(x) = \mathcal{G}(\mathbb{P}(\sigma_{i+1}), x)(u(s''_{i+1} x)) = u(s''_{i+1} x)$$

for all $x = (x_n, \ldots, x_1) \in W_{n+1}$, where y_i, z_i are in (4.6).

A *n*-cochain $u \in C^n(W, \mathcal{G})$ is *normalized* if and only if $\sigma^i u = 0$ for all i. In particular, a 1-cochain u is normalized if $u(1) = 0$; a 2-cochain u is normalized if $u((1, 1), a) = u((1, a), 1) = u((a, 1), 1) = 0$ for all $a \in S$. Under pointwise addition, normalized n-cochains on W constitute a subgroup $C_N^n(W, \mathcal{G})$ of $C^n(W, \mathcal{G})$.

Equation (4.3) yields a coboundary homomorphism δ_n of $C_N^n(W, \mathcal{G})$ into $C_N^{n+1}(W, \mathcal{G})$: if $u \in C_N^n(W, \mathcal{G})$, then

$$\delta_n u = \sum_{0 \le i \le n} (-1)^i \delta^i u \in C^{n+1}(W, \mathcal{G}).$$

By Theorem 4.2.1,

Theorem 4.3.1 *For each $n \ge 0$ there is an isomorphism*

$$H_C^n(S, \mathcal{G}) \cong \operatorname{Ker} \delta_n / \operatorname{Im} \delta_{n-1}$$

which is natural in $\mathcal{G}$.

4.4 Extended Cochains

Extended cochains provide a more convenient description of Calvo-Cegarra cohomology.

4.4.1 Definition

The cochains in Theorem 4.3.1 have finitely many variables but are somewhat unwieldy. More convenient *extended cochains* (called *commutative n-cochains* in [10]) are defined as follows in [10] for dimensions $n \le 4$, using (ordinary) cochains (but no symmetry conditions).

Let S be a commutative monoid and let $\mathcal{G}$ be an abelian group valued functor on S.

Extended 1- and 2-cochains on S with values in $\mathcal{G}$ are (ordinary) normalized 1- and 2-cochains.

An *extended 3-cochain* on S with values in $\mathcal{G}$ is an ordered pair (t, u) of a normalized 3-cochain $t \in C_N^3(S, \mathcal{G})$ and a normalized 2-cochain $u \in C_N^3(S, \mathcal{G})$.

An *extended 4-cochain* on S with values in $\mathcal{G}$ is an ordered triple (q, r, s), where $q \in C_N^4(S, \mathcal{G})$ is a normalized 4-cochain and $r, s \in C_N^3(S, \mathcal{G})$ are normalized 3-cochains.

Under pointwise addition, extended n-cochains on S with values in $\mathcal{G}$ constitute an abelian group $C_e^n(S, \mathcal{G})$ ($n = 1, 2, 3, 4$). For the sake of completeness, $C_e^0(S, \mathcal{G}) = 0$.

The coboundary homomorphism $\delta_n^e : C_e^n(S, \mathcal{G}) \longrightarrow C_e^{n+1}(S, \mathcal{G})$ is defined for $n = 0, 1, 2, 3$ as follows. First, $\delta_0^e = 0$ and $\delta_1^e = \delta_1$ (so that $(\delta_1^e u)(a, b) = u(b)^a - u(ab) + u(a)^b$).

If $v \in C_e^2(S, \mathcal{G})$, then $\delta_2^e v = (t, u) \in C_e^3(S, \mathcal{G})$, where $t = \delta_2 v$:

$$t(a, b, c) = v(b, c)^a - v(ab, c) + v(a, bc) - v(a, b)^c \text{ and}$$

$$u(a, b) = v(b, a) - v(a, b),$$

for all $a, b, c \in S$.

If $w = (t, u) \in C_e^3(S, \mathcal{G})$, then $\delta_3^e w = (q, r, s) \in C_e^4(S, \mathcal{G})$, where $q = \delta_3 t$:

$$q(a, b, c, d) = t(b, c, d)^a - t(ab, c, d)$$
$$+ t(a, bc, d) - t(a, b, cd) + t(a, b, c)^d,$$

$$r(a, b, c) = u(a, c)^b - u(a, bc) + u(a, b)^c$$
$$- t(a, b, c) + t(b, a, c) - t(b, c, a), \text{ and}$$

$$s(a, b, c) = u(b, c)^a - u(ab, c) + u(a, c)^b$$
$$+ t(a, b, c) - t(a, c, b) + t(c, a, b),$$

for all $a, b, c, d \in S$.

It is readily verified that $\delta_{n+1}^e \circ \delta_n^e = 0$, so that

$$Z_e^n(S, \mathcal{G}) = \operatorname{Ker} \delta_n^e \subseteq \operatorname{Im} \delta_{n-1}^e = B_e^n(S, \mathcal{G}).$$

Theorem 4.4.1 ([10]) *If S is a commutative monoid, $\mathcal{G}$ is an abelian group valued functor on S, and $n = 1, 2, 3$, then there is an isomorphism*

$$H_C^n(S, \mathcal{G}) \cong Z_e^n(S, \mathcal{G}) / B_e^n(S, \mathcal{G})$$

which is natural in $\mathcal{G}$.

This is Theorem 3.3 of [10]. The proof, a somewhat lengthy computation, is outlined in some detail in Subsection 3.1 of [10].

4.4.2   Comparison with Symmetric Cohomology

Extended cochains simplify the comparison of the Calvo-Cegarra cohomology of S with its symmetric cohomology. In the diagram

$$
\begin{array}{ccccccccc}
0 & \longrightarrow & C_{NS}^1(S,\,\mathcal{G}) & \xrightarrow{\ \delta_1\ } & C_{NS}^2(S,\,\mathcal{G}) & \xrightarrow{\ \delta_2\ } & C_{NS}^3(S,\,\mathcal{G}) & \xrightarrow{\ \delta_3\ } & C_{NS}^4(S,\,\mathcal{G}) \\
& & \Big\downarrow{\scriptstyle 1} & & \Big\downarrow{\scriptstyle \kappa_2} & & \Big\downarrow{\scriptstyle \kappa_3} & & \Big\downarrow{\scriptstyle \kappa_4} \\
0 & \longrightarrow & C_e^1(S,\,\mathcal{G}) & \xrightarrow{\ \delta_1^e\ } & C_e^2(S,\,\mathcal{G}) & \xrightarrow{\ \delta_2^e\ } & C_e^3(S,\,\mathcal{G}) & \xrightarrow{\ \delta_3^e\ } & C_e^4(S,\,\mathcal{G}),
\end{array}
$$

κ_2 is the inclusion map $C_{NS}^2(S,\,\mathcal{G}) \longrightarrow C_N^2(S,\,\mathcal{G}) = C_e^2(S,\mathcal{G})$,

$$
\kappa_3 t \;=\; (t,0) \quad \text{and} \quad \kappa_4 u \;=\; (u,0,0)
$$

for all $t \in C_{NS}^3(S,\,\mathcal{G})$ and $u \in C_{NS}^4(S,\,\mathcal{G})$. Thus, κ_2, κ_3, and κ_4 are injective.

Since $\delta_1^e = \delta_1$ we have $\kappa_2 \circ \delta_1 = \delta_1^e$. If $v \in C_{NS}^2(S,\,\mathcal{G})$, then v is symmetric and $\delta_2^e v = (\delta_2 v,\, 0) = \kappa_3 \delta_2 v$. Similarly, if $t \in C_{NS}^3(S,\,\mathcal{G})$ is symmetric, then $\kappa_4 \delta_3 t = (\delta_3 t, 0, 0)$, whereas $\delta_3^e \kappa_3 t = \delta_3^e(t, 0) = (q, r, s)$, where $q = \delta_3 t$ and

$$
r(a,b,c) \;=\; -t(a,b,c) + t(b,a,c) - t(b,c,a) \;=\; 0,
$$

$$
s(a,b,c) \;=\; t(a,b,c) - t(a,c,b) + t(c,a,b) \;=\; 0,
$$

for all $a, b, c \in S$, by (S3b). Hence $\delta_3^e \circ \kappa_3 = \kappa_4 \circ \delta_3$ and the diagram (4.4.2) commutes.

Moreover, $B_e^2(S,\,\mathcal{G}) = \delta_1^e\, C_e^1(S,\,\mathcal{G}) = B_{NS}^2(S,\,\mathcal{G})$. If $v \in C_e^2(S,\,\mathcal{G})$, then $\delta_2^e v = 0$ if and only if v is symmetric and $\delta_2 v = 0$; hence $Z_e^2(S,\,\mathcal{G}) = Z_{NS}^2(S,\,\mathcal{G})$. Similarly, $\kappa_3 B_{NS}^3(S,\,\mathcal{G}) = B_{NS}^3(S,\,\mathcal{G})$; and $\kappa_3 Z_{NS}^3(S,\,\mathcal{G}) \subseteq Z_e^3(S,\,\mathcal{G})$. Therefore:

Theorem 4.4.2 ([10]) *If S is a commutative monoid and $\mathcal{G}$ is an abelian group valued functor on S, then there are isomorphisms*

$$
H_C^1(S,\,\mathcal{G}) \cong H_S^1(S,\,\mathcal{G}), \quad H_C^2(S,\,\mathcal{G}) \cong H_S^2(S,\,\mathcal{G})
$$

and an injective homomorphism

$$
H_S^3(S,\,\mathcal{G}) \longrightarrow H_C^3(S,\,\mathcal{G}),
$$

which are natural in $\mathcal{G}$.

4.4.3 An Example

We conclude this section with the example from [10] in which the last map $H_S^3(S, \mathcal{G}) \longrightarrow H_C^3(S, \mathcal{G})$ in Theorem 4.4.2 is not an isomorphism.

Since this example is a group, we begin with the more general case when S is an abelian group. In every abelian group valued functor $\mathcal{G} = (G, \gamma)$ on S, all maps $\gamma_{a,t}$ are then isomorphisms, so that $\mathcal{G}$ is isomorphic to the *constant functor at* $G = G_1$ on S, in which $G_a = G$ and $\gamma_{a,t} = 1_S$ for all $a, t \in S$. Then $H_B^n(S, \mathcal{G})$ is the abelian group cohomology of S with coefficients in G:

$$H_B^2(S, \mathcal{G}) \cong \operatorname{Ext}(S, G) \text{ and } H_B^n(S, \mathcal{G}) = 0 \text{ for all } n \geq 2.$$

Hence it follows from Theorem 3.2.1, 3.2.2, and 3.2.3 that

$$H_S^2(S, \mathcal{G}) \cong \operatorname{Ext}(S, G), \ \ H_S^3(S, \mathcal{G}) = 0, \text{ and } H_S^4(S, \mathcal{G}) = 0.$$

In 'our' example, $S = \{1, x\}$ and $G = \{1, g\}$ are cyclic groups of order 2, and $\mathcal{G}$ is constant at G. Thus S and G are small enough that there are few symmetric n-cochains and even fewer normalized cochains. In fact,

$$C_N^1(S, \mathcal{G}) = \{0, u\}, \ C_N^2(S, \mathcal{G}) = \{0, v\}, \text{ and } C_N^3(S, \mathcal{G}) = \{0, t\},$$

where $u(x) = g$, $v(x, x) = g$, and $t(x, x, x) = g$, so that C_N^1, C_N^2, and C_N^3 are cyclic of order 2.

Note that v is symmetric. Since δv is normalized, by Lemma 3.4.2, and

$$(\delta v)(x, x, x) = v(x, x)^x - v(xx, x) + v(x, xx) - v(x, x)^x = g - g = 0,$$

we have $\delta v = 0$.

On the other hand t is not symmetric, since (S3b) would imply $t(x, x, x) = t(x, x, x) - t(x, x, x) = 0$. Hence $C_{NS}^3(S, \mathcal{G}) = 0$ and Theorem 3.4.2 yields $H_S^3(S, \mathcal{G}) = 0$, as expected. Moreover, for all $a, b, c \in S$,

$$(\delta t)(1, a, b, c) = t(a, b, c)^1 - t(1a, b, c)$$
$$+ t(1, ab, c) - t(1, a, bc) + t(1, a, b)^c = 0,$$

$$(\delta t)(a, 1, b, c) = t(1, b, c)^a - t(a1, b, c)$$
$$+ t(a, 1b, c) - t(a, 1, bc) + t(a, 1, b)^c = 0,$$

$$(\delta t)(a, b, 1, c) = t(b, 1, c)^a - t(ab, 1, c)$$
$$+ t(a, b1, c) - t(a, b, 1c) + t(a, b, 1)^c = 0,$$

$$(\delta t)(a, b, c, 1) = t(b, c, 1)^a - t(ab, c, 1)$$
$$+ t(a, bc, 1) - t(a, b, c1) + t(a, b, c)^1 = 0,$$

$$(\delta t)(x, x, x, x) = t(x, x, x)^x - t(1, x, x)$$
$$+ t(x, 1, x) - t(x, x, 1) + t(x, x, x)^x = g + g = 0,$$

and $\delta t = 0$.

Now we look at the extended cochains of S. By definition,

$$C_e^2(S, \mathcal{G}) = C_N^2(S, \mathcal{G}) = \{0, v\}$$

and

$$C_e^3(S, \mathcal{G}) = C_N^3(S, \mathcal{G}) \times C_N^2(S, \mathcal{G}) = \{(0, 0), (0, v), (t, 0), (t, v)\}.$$

We have $\delta^e v = (q, r)$, where $r = 0$, since v is symmetric, and $q = \delta v = 0$, since q is normalized and

$$q(x, x, x) = v(x, x)^x - v(1, x) + v(x, 1) - v(x, x)^x = 0.$$

Thus, $\delta^e v = (0, 0) = 0$. Hence $B_e^3(S, \mathcal{G}) = 0$.

Next, $\delta^e(0, v) = (q, r, s)$, where $q = \delta 0 = 0$ and

$$r(a, b, c) = v(a, c)^b - v(a, bc) + v(a, b)^c, \text{ and}$$

$$s(a, b, c) = v(b, c)^a - v(ab, c) + v(a, c)^b,$$

for all $a, b, c \in S$. Since v is normalized, $r(1, b, c) = 0$, $r(a, 1, c) = v(a, c) - v(a, c) = 0$, $r(a, b, 1) = -v(a, b) + v(a, b) = 0$, and

$$r(x, x, x) = v(x, x) - v(x, 1) + v(x, x) = g + g = 0.$$

Hence $r = 0$. Similarly, $s(1, b, c) = v(b, c) - v(b, c) = 0$, $s(a, 1, c) = -v(a, c) + v(a, c) = 0$, $s(a, b, 1) = 0$, and

$$s(x, x, x) = v(x, x) - v(1, x) + v(x, x) = g + g = 0;$$

hence $s = 0$. Thus, $(0, v) \in Z_e^3(S, \mathcal{G})$.

Thus it follows from Theorem 4.4.1 that $H_C^3(S, \mathcal{G})$ has at least two elements and is not isomorphic to $H_S^3(S, \mathcal{G})$. (In fact $\delta^e(t, 0) \neq 0$ and $\delta^e(t, v) \neq 0$, so that $H_C^3(S, \mathcal{G})$ has exactly two elements.)

4.5 Properties

The main properties of Calvo-Cegarra cohomology are as follows.

Since the Calvo-Cegarra cohomology of a commutative monoid S is the cohomology of the small category Δ_W, Theorem 4.1.1 yields:

Theorem 4.5.1 *If S is a commutative monoid, then for each short exact sequence*

$$\mathcal{S} \colon 0 \longrightarrow \mathcal{G} \longrightarrow \mathcal{G}' \longrightarrow \mathcal{G}'' \longrightarrow 0$$

of abelian group valued functors on S there is an exact sequence

$$H_C^1(S, \mathcal{G}) \longrightarrow H_C^1(S, \mathcal{G}') \longrightarrow H_C^1(S, \mathcal{G}'') \longrightarrow H_C^2(S, \mathcal{G}) \longrightarrow \cdots$$
$$\cdots H_C^n(S, \mathcal{G}) \longrightarrow H_C^n(S, \mathcal{G}') \longrightarrow H_C^n(S, \mathcal{G}'') \longrightarrow H_C^{n+1}(S, \mathcal{G}) \longrightarrow \cdots$$

which is natural in $\mathcal{S}$.

The other properties follow from Theorem 4.4.2 combined with properties of the symmetric cohomology of S in Chap. 4.

Theorem 4.5.2 *If S is a free commutative monoid, then $H_C^2(S, \mathcal{G}) = 0$ for all $\mathcal{G}$.*

A free commutative monoid is a free commutative semigroup with an identity element adjoined. Hence Theorem 4.5.2 follows from Theorems 4.4.2, 3.4.1, and 3.3.1 (which we saw follows from Theorem 2.3.2). On the other hand it is doubtful that $H_C^3(S)$ vanishes if S is free; otherwise some comparison theorem would probably make H_C^3 isomorphic to H_S^3.

Theorem 4.5.3 *If S is a commutative monoid, then $H_C^2(S, \mathcal{G})$ classifies commutative group coextensions of S by $\mathcal{G}$.*

This follows from Theorems 4.4.2 and 3.1.1.

Chapter 5
The Third Cohomology Group

The third cohomology groups $H^3_S(S, \mathcal{G})$ and $H^3_C(S, \mathcal{G})$ both classify certain monoidal abelian groupoids that may be thought of as extensions of a sort of S by $\mathcal{G}$. This was shown by Calvo-Cervera and Cegarra in [11] and [10]. This chapter is based on [11] and [10], with a fair number of expository changes. In particular, we added concepts that are implicit in [11] and [10], in order to clarify the relationship between monoidal abelian groupoids and 3-cocycles and sharpen the original classification theorems.

5.1 Groupoids

This section contains basic definitions and properties of groupoids and monoidal abelian groupoids.

5.1.1 Groupoids

A *groupoid* is a small category $\mathcal{G}$ in which every morphism is an isomorphism. Groupoids are also called *Brandt groupoids* after Brandt [8].

Every group G can be viewed as a groupoid with one object, in which a morphism is an element of G. Conversely, a groupoid $\mathcal{G}$ with only one object X may be identified with the group $\mathrm{Aut}_{\mathcal{G}}(X)$ of automorphisms of X.

A groupoid $\mathcal{G}$ is *connected* if and only if all objects of $\mathcal{G}$ are isomorphic. The groupoids introduced by Brandt in [8] (as partial semigroup s) have this property. A connected groupoid is equivalent (as a category) to a groupoid with one object.

In any groupoid $\mathcal{G}$, the *connected component* of an object X consists of all objects Y that are isomorphic to X. A *totally disconnected* groupoid is a groupoid in which every connected component consists of only one object (so that $X \cong Y$ if and only if $X = Y$); it may be regarded as a disjoint union of groups.

The connected components of any groupoid constitute a partition of its set of objects. Hence every groupoid is a disjoint union of connected groupoids and is equivalent (as a category) to a totally disconnected groupoid.

An *abelian groupoid* is a groupoid in which all groups $\text{Aut}_G(X)$ are abelian. Thus every abelian groupoid is equivalent (as a category) to a disjoint union of abelian groups, viewed as groupoids with one object.

In [11] and [10], the composite of $u \colon X \longrightarrow Y$ and $v \colon Y \longrightarrow Z$ in an abelian groupoid is written additively, $v + u$. Here we use the familiar notation, $v \circ u$.

5.1.2   Monoidal Groupoids

Abelian groupoids are not complicated enough to provide semigroup theorists with entertainment. This is remedied by providing them with a multiplication.

Generally, a *multiplication* on a category $\mathcal{C}$ is a bifunctor $\cdot \colon \mathcal{C} \times \mathcal{C} \longrightarrow \mathcal{C}$ and sends a pair (X, Y) of objects of $\mathcal{C}$ to an object $XY = X \cdot Y$ and a pair (α, β) of morphisms of $\mathcal{C}$ to a morphism $\alpha \cdot \beta$, so that

$$1_X \cdot 1_Y = 1_{XY} \text{ and } (\alpha \circ \gamma) \cdot (\beta \circ \delta) = (\alpha \cdot \beta) \circ (\gamma \cdot \delta) \tag{5.1}$$

whenever defined. Common notations for $\cdot$ include $\square$ (MacLane [42]), $\otimes$ (Calvo-Cervera and Cegarra [10, 11], among others), and various other symbols.

A *braided monoidal category* $\mathfrak{M}$ is a category endowed with a multiplication for which associativity, identity element, and commutativity hold up to natural isomorphism s: thus, $\mathfrak{M} = (\mathcal{M}, \cdot, I, \mathfrak{a}, \mathfrak{l}, \mathfrak{r}, \mathfrak{c})$ consists of a category $\mathcal{M}$, a multiplication $\cdot$ on $\mathcal{M}$, an identity object I of $\mathcal{M}$, and natural isomorphisms

$$\mathfrak{a} = \mathfrak{a}(X, Y, Z) \colon (XY)Z \longrightarrow X(YZ),$$

$$\mathfrak{l} = \mathfrak{l}(X) \colon IX \longrightarrow X,$$

$$\mathfrak{r} = \mathfrak{r}(X) \colon XI \longrightarrow X, \text{ and}$$

$$\mathfrak{c} = \mathfrak{c}(X, Y) \colon XY \longrightarrow YX,$$

that satisfy the natural compatibility conditions (9.44), (9.45), (9.46), (9.53), and (9.54) in App. 9.4, from [36].

The commutative groupoids defined below, called *strictly symmetric* or *strictly commutative* in [11], are a class of braided monoidal abelian groupoid s: a *commutative monoidal abelian groupoid*

$$\mathfrak{M} = (\mathcal{M}, \cdot, I, \mathfrak{a}, \mathfrak{r}, \mathfrak{c})$$

consists of: an abelian groupoid $\mathcal{M}$, a multiplication $\cdot$ on $\mathcal{M}$, an identity object I, and natural isomorphisms

$$\mathfrak{a} = \mathfrak{a}(X, Y, Z) \colon (XY)Z \longrightarrow X(YZ),$$
$$\mathfrak{r} = \mathfrak{r}(X) \colon XI \longrightarrow X, \quad \text{and}$$
$$\mathfrak{c} = \mathfrak{c}(X, Y) \colon XY \longrightarrow YX,$$

with property (9.44):

$$\mathfrak{a}(X, Y, ZT) \circ \mathfrak{a}(XY, Z, T) \tag{5.2}$$
$$= 1_X \cdot \mathfrak{a}(Y, Z, T) \circ \mathfrak{a}(X, YZ, T) \circ \mathfrak{a}(X, Y, Z) \cdot 1_T;$$

property (9.45):

$$1_X \cdot \mathfrak{r}(Y) \circ \mathfrak{a}(X, I, Y) = \mathfrak{r}(X) \cdot 1_Y; \tag{5.3}$$

property (9.53):

$$\mathfrak{a}(Y, Z, X) \circ \mathfrak{c}(X, YZ) \circ \mathfrak{a}(X, Y, Z) \tag{5.4}$$
$$= 1_Y \cdot \mathfrak{c}(X, Z) \circ \mathfrak{a}(Y, X, Z) \circ \mathfrak{c}(X, Y) \cdot 1_Z;$$

property (9.54):

$$\mathfrak{a}(Z, X, Y)^{-1} \circ \mathfrak{c}(XY, Z) \circ \mathfrak{a}(X, Y, Z)^{-1} \tag{5.5}$$
$$= \mathfrak{c}(X, Z) \cdot 1_Y \circ \mathfrak{a}(X, Z, Y)^{-1} \circ 1_X \cdot \mathfrak{c}(Y, Z);$$

and the *symmetry* property (9.57):

$$\mathfrak{c}(Y, X) = \mathfrak{c}(X, Y)^{-1}; \tag{5.6}$$

for every objects X, Y, Z, T of $\mathcal{M}$. The conditions in [11] are slightly different and more elegant; here we follow Joyal and Street [36].

Property (5.6) makes $\mathfrak{l}$ and (9.55) unnecessary: the latter would read $\mathfrak{l}(X) = \mathfrak{r}(X) \circ \mathfrak{c}(I, X)$. As in App. 9.4 and [42], it also makes $\mathfrak{M}$ a symmetric monoidal category. Hence the Coherence Theorem for symmetric monoidal categories applies to $\mathfrak{M}$: the isomorphisms $\mathfrak{a}, \mathfrak{r}$, and $\mathfrak{c}$ are coherent, so that 'every' diagram that contains only instances of $\mathfrak{a}, \mathfrak{a}^{-1}, \mathfrak{r}, \mathfrak{r}^{-1}, \mathfrak{c}, \mathfrak{c}^{-1}$, and their products, is commutative.

In particular,

$$c(X, X) = 1_{XX} \text{ and} \tag{5.7}$$

$$1_X \cdot t(Y) \circ a(X, Y, I) = t(XY), \tag{5.8}$$

for all objects X, Y of $\mathcal{M}$.

Hiding in plain sight in this definition is a symmetric 3-cocycle, which will be revealed in the next section.

A *morphism* of commutative monoidal abelian groupoids

$$\mathfrak{F} = (\mathcal{F}, \zeta, f) \colon \mathfrak{M} \longrightarrow \mathfrak{M}'$$

from $\mathfrak{M} = (\mathcal{M}, \cdot, I, a, t, c)$ to $\mathfrak{M}' = (\mathcal{M}', \cdot', I', a', t', c')$ (called a *symmetric monoidal functor* in [11], a *braided monoidal functor* in [42] and App. 9.4) consists of a functor $\mathcal{F} \colon \mathcal{M} \longrightarrow \mathcal{M}'$, a natural isomorphism $\zeta(X, Y) \colon \mathcal{F}X \cdot \mathcal{F}Y \longrightarrow \mathcal{F}(XY)$, and an isomorphism $f \colon I' \longrightarrow \mathcal{F}I$, with property (9.49):

$$\mathcal{F}a(X, Y, Z) \circ \zeta(XY, Z) \circ (\zeta(X, Y) \cdot 1_{\mathcal{F}Z}) \tag{5.9}$$

$$= \zeta(X, YZ) \circ (1_{\mathcal{F}X} \cdot \zeta(Y, Z)) \circ a'(\mathcal{F}X, \mathcal{F}Y, \mathcal{F}Z),$$

property (9.50):

$$\mathcal{F}t(X) \circ \zeta(X, I) \circ (1_{\mathcal{F}X} \cdot f) = t'(\mathcal{F}X), \tag{5.10}$$

and property (9.56):

$$\mathcal{F}c(X, Y) \circ \zeta(X, Y) = \zeta(Y, X) \circ c'(\mathcal{F}X, \mathcal{F}Y), \tag{5.11}$$

for all objects X, Y, Z of $\mathcal{M}$. Again, this definition follows [42] rather than the slightly different conditions in [11].

For instance the identity morphism $(\mathcal{F}, \zeta, f)$ on $\mathfrak{M}$ has $\mathcal{F} = 1_{\mathcal{M}}$, $\zeta(X, Y) = 1_{XY}$, and $f = 1_I$.

Morphisms of commutative monoidal abelian groupoids compose as follows: if

$$\mathfrak{F} = (\mathcal{F}, \zeta, f) \colon \mathfrak{M} \longrightarrow \mathfrak{M}' \text{ and } \mathfrak{F}' = (\mathcal{F}', \zeta', f') \colon \mathfrak{M}' \longrightarrow \mathfrak{M}''$$

are morphisms, then $\mathfrak{F}' \circ \mathfrak{F} = (\mathcal{F}'', \zeta'', f'')$, where:

$$\mathcal{F}'' = \mathcal{F}' \circ \mathcal{F}; \tag{5.12}$$

$$\zeta''(X, Y) = \mathcal{F}'\zeta(\mathcal{F}X, \mathcal{F}Y) + \zeta'(\mathcal{F}X, \mathcal{F}Y); \text{ and}$$

$$f'' = \mathcal{F}'f + f'.$$

Commutative monoidal abelian groupoids and their morphisms now constitute a category.

5.1.3 Reduction

Two morphisms $\mathfrak{F} = (\mathcal{F}, \zeta, f)$ and $\mathfrak{F}' = (\mathcal{F}', \zeta', f')$ from $\mathfrak{M}$ to $\mathfrak{M}'$ are *isomorphic* if and only if there exists a natural isomorphism $\theta : \mathcal{F} \longrightarrow \mathcal{F}'$ such that

$$\theta(XY) \circ \zeta(X, Y) = \zeta'(X, Y) \circ \theta(X) \cdot \theta(Y) \text{ and} \tag{5.13}$$

$$\theta(I) \circ f = f',$$

for all objects X, Y of $\mathcal{M}$. The isomorphism θ is called a *symmetric isomorphism* in [11] and is a monoidal natural transformation as defined in [42] and App. 9.4.

Two commutative monoidal abelian groupoids $\mathfrak{M}$ and $\mathfrak{M}'$ are *equivalent* if and only if there exist morphisms $\mathfrak{F} : \mathfrak{M} \longrightarrow \mathfrak{M}'$ and $\mathfrak{F}' : \mathfrak{M}' \longrightarrow \mathfrak{M}$ such that $\mathfrak{F}' \circ \mathfrak{F} \cong 1_{\mathfrak{M}}$ and $\mathfrak{F} \circ \mathfrak{F}' \cong 1_{\mathfrak{M}'}$.

We saw that every groupoid is equivalent to a totally disconnected groupoid. The main result in this section is a similar property of commutative monoidal abelian groupoids.

We call a commutative monoidal abelian groupoid $\mathfrak{M} = (\mathcal{M}, \cdot, I, \mathfrak{a}, \mathfrak{r}, \mathfrak{c})$ *reduced* if and only if $\mathcal{M}$ is totally disconnected, $\mathfrak{r}(X) = 1_X$, and $\mathfrak{c}(X, Y) = 1_{XY}$, for every objects X, Y of $\mathcal{M}$. Reduced commutative monoidal abelian groupoids are called *totally disconnected strictly symmetric abelian groupoids* in [11]. When $\mathcal{M}$ is totally disconnected, the isomorphisms $XI \cong X$ and $XY \cong YX$ imply equalities, $XI = X$ and $XY = YX$; this allows $\mathfrak{r}(X)$ and $\mathfrak{c}(X, Y)$ to equal 1_X and 1_{XY}.

Theorem 5.1.1 *Every commutative monoidal abelian groupoid is equivalent to a reduced commutative monoidal abelian groupoid.*

This is Lemma 3.1 of [11]. Reduction does not obviously extend from groupoids to commutative monoidal abelian groupoids. The proof in [11] involves transferring the given commutative monoidal abelian groupoid structure, $\cdot, I, \mathfrak{a}, \mathfrak{r}, \mathfrak{c}$, to a totally disconnected subgroupoid. Then inspired tinkering produces an equivalence that changes the $\mathfrak{r}$ and $\mathfrak{c}$ maps to identity isomorphisms.

If $\mathfrak{M} = (\mathcal{M}, \cdot, I, \mathfrak{a}, \mathfrak{r}, \mathfrak{c})$ is a reduced commutative monoidal abelian groupoid, then the isomorphisms $\mathfrak{r}$ and $\mathfrak{c}$ do not need to be specified and we simply write $\mathfrak{M} = (\mathcal{M}, \cdot, I, \mathfrak{a})$.

The conditions that define commutative monoidal abelian groupoids become simpler when $\mathfrak{M}$ is reduced. Condition (5.2) remains unchanged, but (5.3), (5.4), and (5.5) become

$$\mathfrak{a}(X, I, Y) = 1_{XY}, \tag{5.14}$$

$$\mathfrak{a}(Y, Z, X) \circ \mathfrak{a}(X, Y, Z) = \mathfrak{a}(Y, Z, X), \quad \text{and} \tag{5.15}$$

$$\mathfrak{a}(Z, X, Y)^{-1} \circ \mathfrak{a}(X, Y, Z)^{-1} = \mathfrak{a}(X, Z, Y)^{-1}, \tag{5.16}$$

for all objects X, Y, Z of $\mathcal{M}$, by (5.1), whereas (5.6) follows from $\mathfrak{c}(X, Y) = 1_{XY}$; property (5.8) becomes

$$\mathfrak{a}(X, Y, I) = 1_{XY}. \tag{5.17}$$

Similarly, conditions (5.10) and (5.11) for morphisms become

$$\zeta(X, I) \circ (1_{\mathcal{F}X} \cdot f) = 1_{\mathcal{F}X} \tag{5.18}$$

$$\zeta(Y, X) = \zeta(X, Y), \tag{5.19}$$

when $\mathfrak{M}$ and $\mathfrak{M}'$ are reduced, and (5.9) remains unchanged.

With their morphisms, reduced commutative monoidal abelian groupoids constitute a full subcategory $\mathcal{R}$ of the category of all commutative monoidal abelian groupoids.

5.1.4   The Base

Inside every reduced commutative monoidal abelian groupoid can be found a commutative monoid and an abelian group valued functor.

Proposition 5.1.1 *Let* $\mathfrak{M} = (\mathcal{M}, \cdot, I, \mathfrak{a})$ *be a reduced commutative monoidal a-belian groupoid. The set* S *of objects of* $\mathcal{M}$ *is a commutative monoid under* $\cdot$, *with identity element* I, *and* $\mathcal{G} = (G, \gamma)$, *where*

$$G_X = \mathrm{Aut}_{\mathcal{M}}(X) \quad \text{and} \quad \gamma_{X,Y} \colon G_X \longrightarrow G_{XY}, \quad g \longmapsto g^Y = g \cdot 1_Y,$$

is an abelian group valued functor on S. *Moreover,*

$$g \cdot h = g^Y \circ h^X$$

for all $g \in G_X$ *and* $h \in G_Y$, *so that* $\mathcal{M}$, $\cdot$, *and* I *are completely determined by* S *and* $\mathcal{G}$.

We call the pair $(S, \mathcal{G})$ in Proposition 5.1.1 the *base* of the reduced commutative monoidal abelian groupoid $\mathfrak{M}$.

Proof *([11])* Since $\mathcal{M}$ is totally disconnected, the isomorphisms $\mathfrak{a} \colon (XY)Z \cong X(YZ)$, $\mathfrak{r} \colon XI \cong X$ and $\mathfrak{c} \colon XY \cong YX$ imply equalities $(XY)Z = X(YZ)$, $XI = X$, and $XY = YX$. Hence S is a commutative monoid.

Next, $G_X = \mathrm{Aut}_{\mathcal{M}}(X)$ is an abelian group (written multiplicatively) since $\mathcal{M}$ is abelian.

The morphisms of $\mathcal{M}$ also constitute a commutative monoid. Indeed the following squares all commute:

$$
\begin{array}{ccc}
(XY)Z \xrightarrow{\mathfrak{a}(X,Y,Z)} X(YZ) & XI = X \xrightarrow{1_X} X & XY \xrightarrow{1_{XY}} YX \\
\downarrow{\scriptstyle(g\cdot h)\cdot k} \qquad \downarrow{\scriptstyle g\cdot(h\cdot k)} & \downarrow{\scriptstyle g\cdot 1_I} \qquad \downarrow{\scriptstyle g} & \downarrow{\scriptstyle g\cdot h} \qquad \downarrow{\scriptstyle h\cdot g} \\
(XY)Z \xrightarrow{\mathfrak{a}(X,Y,Z)} X(YZ) & XI = X \xrightarrow{1_X} X & XY \xrightarrow{1_{XY}} YX
\end{array}
$$

since $\mathfrak{a}$, $\mathfrak{r}$, and $\mathfrak{c}$ are natural transformations. Hence

$$(g\cdot(h\cdot k)) \circ \mathfrak{a}(X,Y,Z) = \mathfrak{a}(X,Y,Z) \circ ((g\cdot h)\cdot k)$$

whenever $g \in G_X$, $h \in G_Y$, and $k \in G_Z$, which in the abelian group G_{XYZ} implies

$$g\cdot(h\cdot k) = (g\cdot h)\cdot k; \tag{5.20}$$

also,

$$g = g \circ 1_X = 1_X \circ (g\cdot 1_I) = g\cdot 1_I, \quad \text{and} \tag{5.21}$$

$$h\cdot g = (h\cdot g) \circ 1_{XY} = 1_{XY} \circ (g\cdot h) = g\cdot h; \tag{5.22}$$

in particular,

$$g \cdot 1_Y = 1_Y \cdot g. \tag{5.23}$$

It now follows from (5.1) that

$$\gamma_{X,Y}: g \mapsto g^Y = g\cdot 1_Y = 1_Y\cdot g \tag{5.24}$$

is a homomorphism of G_X into G_{XY}. In particular,

$$1_X \cdot 1_Y = 1_{XY}. \tag{5.25}$$

Moreover,

$$g^{YZ} = g\cdot 1_{YZ} = g\cdot(1_Y\cdot 1_Z) = (g\cdot 1_Y)\cdot 1_Z = (g^Y)^Z \text{ and}$$

$$g^I = g\cdot 1_I = g,$$

for all $g \in G_X$ and all X, Y, Z, by (5.25), (5.20) and (5.21). Thus $\mathcal{G} = (G, \gamma)$ is an abelian group valued functor on S.

Finally, (5.1) and (5.23) yield

$$g \cdot h \; = \; (g \circ 1_X) \cdot (1_Y \circ h) \; = \; (g \cdot 1_Y) \circ (1_X \cdot h) \; = \; g^Y \circ h^X \qquad (5.26)$$

for all $g \in G_X$ and $h \in G_Y$. Hence $\mathcal{M}$, $\cdot$, and I are completely determined by S and $\mathcal{G}$. Moreover, the commutative monoid of morphisms of $\mathcal{M}$ is isomorphic to the canonical split coextension of S by $\mathcal{G}$. $\qquad\qquad\qquad\qquad\qquad\qquad\square$

In Proposition 5.1.1, $\mathfrak{M}$ 'should be thought of [as] a sort of 2-dimensional twisted crossed product of S by $\mathcal{G}$' [11], or a sort of 2-dimensional split group coextension.

5.2   Symmetric 3-Cocycles

The results in this section link reduced commutative monoidal abelian groupoids and symmetric cohomology.

5.2.1   Cocycle Objects

Writing the abelian groups $G_X = \mathrm{Aut}_{\mathcal{M}}(X)$ additively reveals the hidden cocycles in our (actually, Calvo-Cervera and Cegarra's) groupoids.

Lemma 5.2.1 *If* $\mathfrak{M} = (\mathcal{M}, \cdot, I, \mathfrak{a})$ *is a reduced commutative monoidal abelian groupoid with base* $(S, \mathcal{G})$, *then* $\mathfrak{a}$ *is a normalized symmetric 3-cocycle on* S *with values in* $\mathcal{G}$.

Proof We have $\mathfrak{a}(X, Y, Z) \in \mathrm{Aut}_{\mathcal{M}}(XYZ) = G_{XYZ}$ for all $X, Y, Z \in S$, so that $\mathfrak{a} \in C^3(S, \mathcal{G})$. In the additive notation, (5.15) and (5.16) read

$$\mathfrak{a}(Y, Z, X) + \mathfrak{a}(X, Y, Z) \; = \; \mathfrak{a}(Y, X, Z) \; \text{ and}$$

$$-\mathfrak{a}(Z, X, Y) - \mathfrak{a}(X, Y, Z) \; = \; -\mathfrak{a}(X, Z, Y),$$

for all $X, Y, Z \in S$. Each condition is equivalent to (S3b). Therefore $\mathfrak{a}$ is symmetric. Similarly, (5.2) reads

$$\mathfrak{a}(X, Y, ZT) + \mathfrak{a}(XY, Z, T) \; = \; \mathfrak{a}(Y, Z, T)^X + \mathfrak{a}(X, YZ, T) + \mathfrak{a}(X, Y, Z)^T$$

for all $X, Y, Z, T \in S$, using (5.21), and states that $\delta \mathfrak{a} = 0$, so that $\mathfrak{a}$ is a 3-cocycle. Finally, $\mathfrak{a}$ is normalized, by (5.17). $\qquad\qquad\qquad\qquad\qquad\qquad\square$

We now bundle the base of $\mathfrak{M}$ and its 3-cocycle $\mathfrak{a}$ into one *symmetric 3-cocycle object* $\mathrm{T} = (S, \mathcal{G}, t)$ that consists of a commutative monoid S, an abelian group valued functor $\mathcal{G}$ on S, and a normalized symmetric 3-cocycle $t \in Z^3_{NS}(S, \mathcal{G})$.

Combining Lemma 5.2.1 and Proposition 5.1.1 yields:

Proposition 5.2.1 *If $\mathfrak{M} = (\mathcal{M}, \cdot, I, \mathfrak{a})$ is a reduced commutative monoidal abelian groupoid with base $(S, \mathcal{G})$, then*

$$\mathbb{T}(\mathfrak{M}) = (S, \mathcal{G}, \mathfrak{a})$$

is a symmetric 3-cocycle object. Moreover, $\mathbb{T}(\mathfrak{M}) = \mathbb{T}(\mathfrak{M}')$ implies $\mathfrak{M} = \mathfrak{M}'$.

Conversely:

Proposition 5.2.2 *For every symmetric 3-cocycle object T there exists a reduced commutative monoidal abelian groupoid $\mathfrak{M}$ such that $\mathbb{T}(\mathfrak{M}) = T$.*

Proof ([11]) Let $T = (S, \mathcal{G}, t)$ be a symmetric 3-cocycle object. Let $\mathcal{M}$ be the disjoint union of all abelian groups G_a: the totally disconnected groupoid in which an object is an element of S, a morphism $a \longrightarrow a$ is an element of G_a, and composition in each G_a is by addition in G_a, $u \circ v = u + v$. Let

$$a \cdot b = ab \text{ and } g \cdot h = g^b + h^a \in G_{ab}$$

for all $a, b \in S$, $g \in G_a$, and $h \in G_b$, as in (5.26). It is straightforward that $\cdot$ is a bifunctor. Moreover,

$$(g \cdot h) \cdot k = g \cdot (h \cdot k)$$

for all $g \in G_a$, $h \in G_b$, and $k \in G_c$, which in the abelian group G_{abc} implies

$$g \cdot (h \cdot k) + t(a, b, c) = t(a, b, c) + (g \cdot h) \cdot k.$$

Hence $t(a, b, c): (ab)c \longrightarrow a(bc)$ is natural in a, b, c.

Finally, let I be the identity element of S. Then $\mathfrak{M} = (\mathcal{M}, \cdot, I, t)$ is a reduced commutative monoidal abelian groupoid: (5.2) holds since t is a cocycle; (5.15) and (5.16) hold since t is symmetric; and (5.14) holds since t is normalized. It is clear that $\mathbb{T}(\mathfrak{M}) = T$. $\square$

5.2.2 Morphisms

Next we extend $\mathbb{T}$ to a functor. This requires morphisms of symmetric 3-cocycle objects, whose definition uses the following constructions.

Every monoid homomorphism $\varphi: S \longrightarrow S'$ induces a functor $\mathcal{H}(\varphi)$ of $\mathcal{H}(S)$ into $\mathcal{H}(S')$, defined by

$$\mathcal{H}(\varphi)(a) = \varphi(a) \text{ and } \mathcal{H}(\varphi)(a, b) = (\varphi(a), \varphi(b))$$

for all $a, b \in S$. If $\varphi: S \longrightarrow S'$ and $\varphi': S' \longrightarrow S''$ are monoid homomorphisms, then $\mathcal{H}(\varphi' \circ \varphi) = \mathcal{H}(\varphi') \circ \mathcal{H}(\varphi)$. Hence every abelian group valued functor $\mathcal{G}' = (G', \gamma')$ on S' induces an abelian group valued functor $\mathcal{G}'_\varphi = (G, \gamma)$ on S:

$$\mathcal{G}'_\varphi = \mathcal{G}' \circ \mathcal{H}(\varphi),$$

in which

$$(\mathcal{G}'_\varphi)_a = G'_{\varphi(a)} = G_a \text{ and } \gamma'_{\varphi(a), \varphi(b)} = \gamma_{a,b}, \text{ so that } g^{\varphi(b)} = g^b,$$

for all $a, b \in S$ and $g \in G_a$.

Let $\varphi: S \longrightarrow S'$ be a monoid homomorphism. Every natural transformation $\sigma: \mathcal{G} \longrightarrow \mathcal{G}'_\varphi$ induces for each $n > 0$ a homomorphism

$$\sigma_*: C^n(S', \mathcal{G}') \longrightarrow C^n(S, \mathcal{G}'_\varphi),$$

$$(\sigma_* u)(a_1, a_2, \ldots, a_n) = \sigma(a)\, u(a_1, a_2, \ldots, a_n) \in G'_{\varphi(a)}, \qquad (5.27)$$

for all $a_1, a_2, \ldots, a_n \in S$ and $u \in C^n(S', \mathcal{G}')$, where $a = a_1 a_2 \cdots a_n$. Then $(\sigma' \sigma)_* = \sigma'_* \sigma_*$; if $n \leq 4$, then $\sigma_* \delta u = \delta \sigma_* u$ for all $u \in C^n(S', \mathcal{G}')$.

If $n \leq 3$ and u is a normalized symmetric n-cocycle, then so is $\sigma_* u$, so that σ_* induces a homomorphism $\sigma_*: Z^n_{NS}(S', \mathcal{G}') \longrightarrow Z^n_{NS}(S, \mathcal{G}'_\varphi)$.

The monoid homomorphism $\varphi: S \longrightarrow S'$ also induces a homomorphism

$$\varphi^*: C^n(S', \mathcal{G}') \longrightarrow C^n(S, \mathcal{G}'_\varphi),$$

$$(\varphi^* u)(a_1, a_2, \ldots, a_n) = u(\varphi(a_1), \varphi(a_2), \ldots, \varphi(a_n)) \in G'_{\varphi(a)}, \qquad (5.28)$$

for all $a_1, a_2, \ldots, a_n \in S$ and $u' \in C^n(S', \mathcal{G}')$, where $a = a_1 a_2 \cdots a_n$. Then $(\psi \varphi)^* = \varphi^* \psi^*$; if $n \leq 4$, then $\varphi^* \delta u = \delta \varphi^* u$ for all $u \in C^n(S', \mathcal{G}')$.

If $n \leq 3$ and u is a normalized symmetric 3-cocycle, then so is $\varphi^* u'$, so that φ^* induces a homomorphism $\varphi^*: Z^3_{NS}(S', \mathcal{G}') \longrightarrow Z^3_{NS}(S, \mathcal{G}'_\varphi)$.

Finally, let $\varphi: S \longrightarrow S'$ and $\psi: S' \longrightarrow S''$ be monoid homomorphisms and let $\tau: \mathcal{G}' \longrightarrow \mathcal{G}''_\psi$ be a natural transformation. In addition to $\tau_*: C^n(S', \mathcal{G}') \longrightarrow C^n(S', \mathcal{G}''_\psi)$, τ also induces a natural transformation

$$\tau^\varphi = \tau \circ \varphi: \mathcal{G}'_\varphi \longrightarrow (\mathcal{G}''_\psi)_\varphi$$

that assigns

$$\tau^\varphi(a) = \tau(\varphi(a)): G'_{\varphi(a)} \longrightarrow G''_{\psi(\varphi(a))}$$

to each $a \in S$. Then $\tau_*^\varphi \circ \varphi^* = \varphi^* \circ \tau_*$:

$$
\begin{array}{ccc}
C^n(S', \mathcal{G}') & \xrightarrow{\varphi^*} & C^n(S, \mathcal{G}'_\varphi) \\
{\scriptstyle \tau_*} \downarrow & & \downarrow {\scriptstyle \tau_*^\varphi} \\
C^n(S', \mathcal{G}''_\psi) & \xrightarrow{\varphi^*} & C^n(S, \mathcal{G}''_{\psi \circ \varphi})
\end{array}
$$

We can now define morphisms of symmetric 3-cocycle objects, so that symmetric 3-cocycle objects become the objects of a category $\mathcal{T}$. A *morphism* $\Phi = (\varphi, \sigma, u, f)$ in $\mathcal{T}$ from $T = (S, \mathcal{G}, t)$ to $T' = (S', \mathcal{G}', t')$ consists of a monoid homomorphism $\varphi: S \longrightarrow S'$, a natural transformation $\sigma: \mathcal{G} \longrightarrow \mathcal{G}'_\varphi$, a 2-cochain $u \in C^2_S(S, \mathcal{G}'_\varphi)$, and an element f of G'_1, such that

$$\varphi^* t' = \sigma_* t - \delta u \text{ and} \tag{5.29}$$

$$u(a, 1) = -f^{\varphi(a)}, \tag{5.30}$$

for all $a \in S$.

For instance, the identity morphism on $T = (S, \mathcal{G}, t)$ is (φ, σ, u, f), where $\varphi(a) = a$, $\sigma(a)g = g$, $u(a, b) = 0$, and $f = 0$, for all $a, b \in S$ and $g \in G_a$.

Morphisms of symmetric 3-cocycle objects are composed as follows. If $(\varphi, \sigma, u, f): T \longrightarrow T'$ and $(\psi, \tau, v, g): T' \longrightarrow T''$ are morphisms in $\mathcal{T}$, then

$$(\psi, \tau, v, g) \circ (\varphi, \sigma, u, f) = (\psi \circ \varphi, \tau^\varphi \circ \sigma, \tau_*^\varphi u + \varphi^* v, \tau(1)f + g). \tag{5.31}$$

Composition is well defined: in (5.31), we have $\varphi^* t' = \sigma_* t - \delta u$ and $\psi^* t'' = \tau_* t' - \delta v$, so that

$$
\begin{aligned}
(\psi \circ \varphi)^* t'' &= \varphi^* \psi^* t'' = \varphi^*(\tau_* t' - \delta v) \\
&= \tau_*^\varphi \varphi^* t' - \varphi^* \delta v = \tau_*^\varphi(\sigma_* t - \delta u) - \varphi^* \delta v \\
&= \tau_*^\varphi \sigma_* t - \delta(\tau_*^\varphi u + \varphi^* v);
\end{aligned}
$$

also, $u(a, 1) = -f^{\varphi(a)}$, $v(b, 1) = -g^{\psi(b)}$, so that, for all $a \in S$,

$$
\begin{aligned}
(\tau_*^\varphi u + \varphi^* v)(a, 1) &= (\tau_*^\varphi) u(a, 1) + (\varphi^* v)(a, 1) \\
&= -\tau(\varphi(a)) f^{\varphi(a)} + v(\varphi(a), 1) = -\tau(1) f^{\psi(\varphi(a))} - g^{\psi(\varphi(a))} \\
&= -(\tau(1) f + g)^{\psi(\varphi(a))},
\end{aligned}
$$

where $\tau(\varphi(a)) \circ \gamma'_{1,\varphi(a)} = \gamma''_{1,\psi(\varphi(a))} \circ \tau(1)$ since τ is a natural transformation.

Two straightforward results now complete the construction of $\mathbb{T}$ as a functor.

Lemma 5.2.2 *Let $\mathfrak{F} = (\mathcal{F}, \zeta, f)$ be a morphism in $\mathcal{R}$ from $\mathfrak{M}$ to $\mathfrak{M}'$. For every object X and morphism $g: x \longrightarrow x$ of $\mathcal{M}$ let*

$$\varphi(X) = \mathcal{F}X \quad and \quad \sigma(X)\,g = \mathcal{F}g.$$

Then $\mathbb{T}(\mathfrak{F}) = (\varphi, \sigma, \zeta, f)$ is a morphism in $\mathcal{T}$ from $\mathbb{T}(\mathfrak{M})$ to $\mathbb{T}(\mathfrak{M}')$. Moreover, if $\mathfrak{F}'$ is a morphism from $\mathfrak{M}'$ to $\mathfrak{M}''$, then $\mathbb{T}(\mathfrak{F}' \circ \mathfrak{F}) = \mathbb{T}(\mathfrak{F}') \circ \mathbb{T}(\mathfrak{F})$; if $\mathfrak{F}$ and $\mathfrak{F}'$ are morphisms from $\mathfrak{M}$ to $\mathfrak{M}'$, then $\mathbb{T}(\mathfrak{F}) = \mathbb{T}(\mathfrak{F}')$ implies $\mathfrak{F} = \mathfrak{F}'$.

Proof First, $\mathcal{F}$ is completely determined by φ and σ; hence $\mathbb{T}(\mathfrak{F}) = \mathbb{T}(\mathfrak{F}')$ implies $\mathfrak{F} = \mathfrak{F}'$.

Let $\mathfrak{M} = (\mathcal{M}, \cdot, I, \mathfrak{a})$ and $\mathfrak{M}' = (\mathcal{M}', \cdot', I', \mathfrak{a}')$ be reduced, let $\mathbb{T}(\mathfrak{M}) = (S, \mathcal{G}, t)$, and let $\mathbb{T}(\mathfrak{M}') = (S', \mathcal{G}', t')$. Since $\mathfrak{M}'$ is reduced, the isomorphisms

$$\zeta(X, Y): \mathcal{F}X \cdot \mathcal{F}Y \longrightarrow \mathcal{F}(XY) \quad and \quad f: I' \longrightarrow \mathcal{F}I$$

imply $\mathcal{F}X \cdot \mathcal{F}Y = \mathcal{F}(XY)$ and $I' = \mathcal{F}I$, so that φ is a monoid homomorphism of S into S'. By (5.28),

$$(\varphi^* t')(X, Y, Z) = t\,(\mathcal{F}X, \mathcal{F}Y, \mathcal{F}Z). \tag{5.32}$$

If $g \in G_X = \mathrm{Aut}_{\mathcal{M}}(X)$, then $\mathcal{F}g \in \mathrm{Aut}_{\mathcal{M}'}(\mathcal{F}X) = G'_{\varphi(X)}$; thus $\mathcal{F}$ induces a homomorphism $\sigma(X): g \longmapsto \mathcal{F}g$ of G_X into $G'_{\varphi(X)}$. The naturality of σ follows from that of ζ: for any $g: X \longrightarrow X$ and $h: Y \longrightarrow Y$ in $\mathcal{M}$, the square

$$\begin{array}{ccc} \mathcal{F}X \cdot \mathcal{F}Y & \xrightarrow{\zeta(X,Y)} & \mathcal{F}(XY) \\ {\scriptstyle \mathcal{F}g \cdot \mathcal{F}h}\downarrow & & \downarrow{\scriptstyle \mathcal{F}(g \cdot h)} \\ \mathcal{F}X \cdot \mathcal{F}Y & \xrightarrow{\zeta(X,Y)} & \mathcal{F}(XY) \end{array}$$

commutes, since ζ is natural, so that $\mathcal{F}(g \cdot h) \circ \zeta(X, Y) = \zeta(X, Y) \circ (\mathcal{F}g \cdot \mathcal{F}y)$, which in the abelian group $G_{\mathcal{F}(XY)}$ implies

$$\mathcal{F}(g \cdot h) = \mathcal{F}g \cdot \mathcal{F}h.$$

Since $h \longmapsto \mathcal{F}h$ is a homomorphism we also have $\mathcal{F}1_Y = 1_{\mathcal{F}Y}$. Hence

$$\sigma(XY)(g^Y) = \mathcal{F}(g \cdot 1_Y) = \mathcal{F}g \cdot 1_{\mathcal{F}Y} = (\mathcal{F}g)^{\mathcal{F}Y},$$

the square

$$\begin{array}{ccc} G_X & \xrightarrow{\sigma(X)} & G'_{\mathcal{F}X} \\ {\scriptstyle \gamma}\downarrow & & \downarrow{\scriptstyle \gamma'} \\ G_{XY} & \xrightarrow{\sigma(XY)} & G'_{\mathcal{F}(XY)} \end{array}$$

commutes, and σ is a natural transformation $\mathcal{G} \longrightarrow \mathcal{G}'_\varphi$. By (5.27),

$$(\sigma_* t)(X, Y, Z) = \sigma(XYZ)\,t(X, Y, Z) = \mathcal{F}t\,(X, Y, Z)\,. \tag{5.33}$$

Since $c(X, Y) = 1_{XY}$ and $c' = 1_{\mathcal{F}X} \cdot {}_{\mathcal{F}Y} = 1_{\mathcal{F}(XY)}$, (5.11) reads

$$\mathcal{F}1_{XY} \circ \zeta(X, Y) = \zeta(Y, X) \circ 1_{\mathcal{F}(XY)}\,,$$

which in the abelian group $G'_{\mathcal{F}(XY)}$ implies $\zeta(Y, X) = \zeta(X, Y)$, for all $X, Y \in S$. Thus ζ is a symmetric 2-cochain, $\zeta \in C_S^2(S, \mathcal{G}'_\varphi)$. Equation (5.9) now yields

$$\mathcal{F}t(X, Y, Z) \circ \zeta(XY, Z) \circ \zeta(X, Y)^Z$$
$$= \zeta(X, YZ) \circ \zeta(Y, Z)^X \circ t'(\mathcal{F}X, \mathcal{F}Y, \mathcal{F}Z),$$

so that (5.32) and (5.33) yield

$$\varphi^* t' = \sigma_* t - \delta\zeta\,.$$

Finally, f is an automorphism of $I' = \mathcal{F}I$, so $f \in G'_1$, and (5.10) yields

$$1_{\mathcal{F}X} \circ \zeta(X, I) \circ (1_{\mathcal{F}X} \cdot f) = 1_{\mathcal{F}X}$$

and $\zeta(X, I) = -f^{\varphi(X)}$, for all $X \in S$. Thus $\mathbb{T}(\mathfrak{F}) = (\varphi, \sigma, \zeta, f)$ is a morphism from $\mathbb{T}(\mathfrak{M})$ to $\mathbb{T}(\mathfrak{M}')$.

Now let $\mathfrak{F}' = (\mathcal{F}', \zeta', f')$ be a morphism from $\mathfrak{M}$ to $\mathfrak{M}'$. Then $\mathbb{T}(\mathfrak{F}') = (\psi, \tau, \zeta', f')$, where $\psi(X') = \mathcal{F}'X'$, $\tau(X')\,g' = \mathcal{F}'g'$, for all $X' \in S'$ and $g' \in G'_{\mathcal{F}'X'}$, and

$$\mathbb{T}(\mathfrak{F}') \circ \mathbb{T}(\mathfrak{F}) = (\psi \circ \varphi, \tau^\varphi \circ \sigma, \tau_*^\varphi \zeta + \varphi^* \zeta', \tau(1)f + f')\,,$$

by (5.31). On the other hand, $\mathbb{T}(\mathfrak{F}' \circ \mathfrak{F}) = (\chi, \omega, \zeta'', f'')$, where

$$\chi(X) = \mathcal{F}'\mathcal{F}X, \quad \omega(X)\,g = \mathcal{F}'\mathcal{F}g,$$
$$\zeta''(X, Y) = \mathcal{F}'\,\zeta(\mathcal{F}X, \mathcal{F}Y) + \zeta'(\mathcal{F}'\mathcal{F}X, \mathcal{F}'\mathcal{F}Y), \quad \text{and}$$
$$f'' = \mathcal{F}'f + f'.$$

for all $X, Y \in S$ and $g \in G_{\mathcal{F}X}$, by (5.12). Then $\chi = \psi \circ \varphi$,

$$\tau^\varphi(X)\,\sigma(X)\,g = \tau^\varphi(X)\,\mathcal{F}g = \tau(\mathcal{F}X)\,\mathcal{F}g = \mathcal{F}'\mathcal{F}g,$$

so that $\omega = \tau^\varphi \circ \sigma$; $\zeta'' = \tau_*^\varphi \zeta + \varphi^* \zeta'$; and $f'' = \mathcal{F}'f + f' = \tau(1)f + f'$. Thus $\mathbb{T}(\mathfrak{F}' \circ \mathfrak{F}) = \mathbb{T}(\mathfrak{F}') \circ \mathbb{T}(\mathfrak{F})$. $\square$

Conversely:

Proposition 5.2.3 *Let* $T = \mathbb{T}(\mathfrak{M})$ *and* $T' = \mathbb{T}(\mathfrak{M}')$. *For every morphism* Φ *from* T *to* T' *there exists a morphism* $\mathfrak{F}$ *from* $\mathfrak{M}$ *to* $\mathfrak{M}'$ *such that* $\mathbb{T}(\mathfrak{F}) = \Phi$.

Proof Let $\mathfrak{M} = (\mathcal{M}, \cdot, I, \mathfrak{a})$, $\mathfrak{M}' = (\mathcal{M}', \cdot', I', \mathfrak{a}')$, $T = (S, \mathcal{G}, t)$, $T' = (S', \mathcal{G}', t')$, and let $\Phi = (\varphi, \sigma, u, f)$ be a morphism from T to T'. For every $a \in S$ and $g \in G_a$, define, as we must,

$$\mathfrak{F}a = \varphi(a) \text{ and } \mathfrak{F}g = \sigma(a)\,g.$$

We show that $\mathfrak{F} = (\mathfrak{F}, u, f)$ is a morphism from $\mathfrak{M}$ to $\mathfrak{M}'$. Then $\mathbb{T}(\mathfrak{F}) = \Phi$.

First, $\mathfrak{F}$ is a functor from $\mathcal{M}$ to $\mathcal{M}'$, since each $\sigma(a)$ is a homomorphism.

For each $a, b \in S$, $u(a, b) \in G'_{ab}$ is a morphism in $\mathcal{M}'$ from $\mathfrak{F}a \cdot \mathfrak{F}b$ to $\mathfrak{F}(ab)$. The naturality of u follows from that of σ: the square

$$
\begin{array}{ccc}
G_a & \xrightarrow{\ \gamma\ } & G_{ab} \\
{\scriptstyle\sigma}\downarrow & & \downarrow{\scriptstyle\sigma} \\
G'_a & \xrightarrow{\ \gamma'\ } & G'_{ab}
\end{array}
$$

commutes, so that $\sigma(ab)\,g^b = (\sigma(a)\,g)^b$ for all $g \in G_a$; similarly, $\sigma(ab)\,h^a = (\sigma(b)\,h)^a$ for all $h \in G_b$. Hence

$$\mathfrak{F}(g \cdot h) = \sigma(ab)(g^b \circ h^a) = (\sigma(a)\,g)^b \circ (\sigma(b)\,h)^a = \mathfrak{F}g \cdot \mathfrak{F}h$$

for all $g \in G_a$ and $h \in G_b$. Then the square

$$
\begin{array}{ccc}
\mathfrak{F}a \cdot \mathfrak{F}b & \xrightarrow{\ u(a,b)\ } & \mathfrak{F}(ab) \\
{\scriptstyle\mathfrak{F}g \cdot \mathfrak{F}h}\downarrow & & \downarrow{\scriptstyle\mathfrak{F}(g \cdot h)} \\
\mathfrak{F}a \cdot \mathfrak{F}b & \xrightarrow{\ u(a,b)\ } & \mathfrak{F}(ab)
\end{array}
$$

commutes, since G'_{ab} is abelian. Thus u is natural.

Finally, (5.9) follows from $\varphi^* t' = \sigma_* t - \delta u$; (5.18) follows from $u(a, 1) = -f^{\varphi(a)}$; and (5.19) holds since u is symmetric.   □

It follows from Lemma 5.2.1 and Proposition 5.2.1, 5.2.2, 5.2.3 that $\mathbb{T}$ is an isomorphism of categories. Hence

Theorem 5.2.1 *The category* $\mathcal{R}$ *of reduced commutative monoidal abelian groupoids is isomorphic to the category* $\mathcal{T}$ *of symmetric 3-cocycle objects.*

5.3 Classification

We now turn to the Classification Theorems.

5.3.1 Isomorphisms

Recall that two morphisms $\mathfrak{F} = (\mathcal{F}, \zeta, f)$ and $\mathfrak{F}' = (\mathcal{F}', \zeta', f')$ of commutative monoidal abelian groupoids from $\mathfrak{M}$ to $\mathfrak{M}'$ are *isomorphic* if and only if there exists a natural isomorphism $\theta: \mathcal{F} \longrightarrow \mathcal{F}'$ such that (5.13):

$$\theta(XY) + \zeta(X, Y) = \zeta'(X, Y) + \theta(X) \cdot \theta(Y) \text{ and}$$
$$\theta(I) + f = f'$$

holds for all objects X, Y of $\mathcal{M}$. In particular this definition applies to morphisms in $\mathcal{R}$.

The corresponding definition for morphisms of $\mathcal{T}$ is: two coterminal morphisms $\Phi = (\varphi, \sigma, u, f)$ and $\Phi' = (\varphi', \sigma', u', f')$ are *isomorphic* if and only if $\varphi = \varphi'$, $\sigma = \sigma'$, and there exists a 1-cochain $p \in C^1(S, \mathcal{G}'_\varphi)$ such that

$$u' = u - \delta p \text{ and } f' = f + p(1). \tag{5.34}$$

Proposition 5.3.1 *Two coterminal morphisms $\mathfrak{F}$ and $\mathfrak{F}'$ of $\mathcal{R}$ are isomorphic if and only if $\mathbb{T}(\mathfrak{F})$ and $\mathbb{T}(\mathfrak{F}')$ are isomorphic.*

Proof Assume that $\mathfrak{F} = (\mathcal{F}, \zeta, f)$ and $\mathfrak{F}' = (\mathcal{F}', \zeta', f')$ are isomorphic morphisms from $\mathfrak{M} = (\mathcal{M}, \cdot, I, \mathfrak{a})$ to $\mathfrak{M}' = (\mathcal{M}', \cdot', I', \mathfrak{a}')$, where $\mathfrak{M}$ and $\mathfrak{M}'$ are reduced. Then there is a natural isomorphism $\theta: \mathcal{F} \longrightarrow \mathcal{F}'$ such that (5.13) holds for all objects X, Y of $\mathcal{M}$.

Since $\mathfrak{M}'$ is reduced, the isomorphism $\theta(X): \mathcal{F}X \longrightarrow \mathcal{F}'X$ implies $\mathcal{F}X = \mathcal{F}'X$, for every object X of $\mathcal{M}$. Since θ is natural the square

$$
\begin{array}{ccc}
\mathcal{F}X & \xrightarrow{\theta} & \mathcal{F}'X \\
{\scriptstyle \mathcal{F}g} \downarrow & & \downarrow {\scriptstyle \mathcal{F}'g} \\
\mathcal{F}X & \xrightarrow{\theta} & \mathcal{F}'X
\end{array}
$$

commutes for every morphism $g: X \longrightarrow X$ of $\mathcal{M}$, so that

$$\mathcal{F}'g + \theta(X) = \theta(X) + \mathcal{F}g.$$

In the abelian group $\mathrm{Aut}_{\mathcal{M}'}(\mathcal{F}'X)$ this implies $\mathcal{F}'g = \mathcal{F}g$. Hence $\mathcal{F}' = \mathcal{F}$.

Let $T = \mathbb{T}(\mathfrak{M}) = (S, \mathcal{G}, t)$, $T' = \mathbb{T}(\mathfrak{M}') = (S', \mathcal{G}', t')$, $\Phi = \mathbb{T}(\mathfrak{F}) = (\varphi, \sigma, u, f)$, and $\Phi' = \mathbb{T}(\mathfrak{F}') = (\varphi', \sigma', u', f')$. Then $\mathfrak{F} = \mathfrak{F}'$ implies $\varphi = \varphi'$ and $\sigma = \sigma'$. Also, $\theta(X) \in \mathrm{Aut}_{\mathcal{M}'}(\mathfrak{F}X) = G'_{\mathfrak{F}X}$, so that $\theta \in C^1(S, \mathcal{G}'_\varphi)$. By (5.13),

$$\theta(1) + f = f' \quad \text{and} \quad \theta(XY) + u(X, Y) = u'(X, Y) + \theta(X)^Y + \theta(Y)^X,$$

so that $u' = u - \delta\theta$ and (5.34) holds. Hence Φ and Φ' are isomorphic.

The converse is straightforward. If Φ and Φ' are isomorphic, then $\varphi = \varphi'$, $\sigma = \sigma'$, and $\mathfrak{F} = \mathfrak{F}'$. Moreover, $p(X) \in G'_{\mathfrak{F}X} = \mathrm{Aut}_{\mathcal{M}}(\mathfrak{F}X)$, since $p \in C^1(S, \mathcal{G}'_\varphi)$, and the isomorphism $p(X)\colon \mathfrak{F}X \longrightarrow \mathfrak{F}X$ is natural in X since $G'_{\mathfrak{F}X}$ is abelian. Then (5.34) yields (5.13). $\qquad \square$

We note some useful properties of isomorphism.

Lemma 5.3.1 *Let* Φ, Φ', Ψ, *and* Ψ' *be morphisms of* $\mathcal{T}$. *If* $\Psi \circ \Phi$ *is defined,* Φ *and* Φ' *are isomorphic, and* Ψ *and* Ψ' *are isomorphic, then* $\Psi \circ \Phi$ *and* $\Psi' \circ \Phi'$ *are isomorphic.*

Proof Let $\Phi = (\varphi, \sigma, u, f)\colon (S, \mathcal{G}, t) \longrightarrow (S', \mathcal{G}', t')$ and $\Psi = (\psi, \tau, v, g)\colon (S', \mathcal{G}', t') \longrightarrow (S'', \mathcal{G}'', t'')$, so that

$$\Psi \circ \Phi = (\psi \circ \varphi, \tau^\varphi \circ \sigma, \tau^\varphi_* u + \varphi^* v, \tau(1)f + g);$$

and let $\Phi' = (\varphi', \sigma', u, f)$ and $\Psi' = (\psi', \tau', v', g')$, so that

$$\Psi' \circ \Phi' = (\psi' \circ \varphi', \tau'^{\varphi'} \circ \sigma', \tau'^{\varphi'}_* u' + \varphi'^* v', \tau'(1)f' + g').$$

By the hypothesis, $\varphi' = \varphi$, $\psi' = \psi$, $\sigma' = \sigma$, $\tau' = \tau$, and there exist 1-cochains $p \in C^1(S, \mathcal{G}'_\varphi)$ and $q \in C^1(S, \mathcal{G}''_\psi)$ such that

$$u' = u - \delta p, \quad v' = v - \delta q, \quad f' = f + p(1), \quad \text{and} \quad g' = g + q(1).$$

Hence $\psi' \circ \varphi' = \psi \circ \varphi$, $\tau'^{\varphi'} \circ \sigma' = \tau^\varphi \circ \sigma$, and

$$\tau'^{\varphi'}_* u' + \varphi'^* v' = \tau^\varphi_* u' + \varphi^* v' = \tau^\varphi_* u + \varphi^* v - \tau^\varphi_* \delta p - \varphi^* \delta q$$
$$= \tau^\varphi_* u + \varphi^* v - \delta \tau^\varphi_* p - \delta \varphi^* q = \tau^\varphi_* u + \varphi^* v - \delta r,$$

where $r = \tau^\varphi_* p + \varphi^* q \in C^1(S, \mathcal{G}''_{\psi \circ \varphi})$. Moreover,

$$\tau'(1)f' + g' = \tau(1)(f + p(1))) + g + q(1) = \tau(1)f + g + r(1),$$

since $\varphi(1) = 1$. Thus $\Psi \circ \Phi$ and $\Psi' \circ \Phi'$ are isomorphic. $\qquad \square$

A morphism $\Phi = (\varphi, \sigma, u, f)$ in $\mathcal{T}$ is *normalized* if and only if $u(a, 1) = 0$ for all a; equivalently, by (5.30), if $f = 0$.

Proposition 5.3.2 *Every morphism of symmetric 3-cocycle objects is isomorphic to a normalized morphism.*

Proof Let $\Phi = (\varphi, \sigma, u, f)\colon (S, \mathcal{G}, t) \longrightarrow (S', \mathcal{G}', t')$. Since t and t' are normalized, then so is $\delta u = \varphi^* t' - \sigma_* t$. By Lemma 3.4.4, $\delta u = \delta v$ for some normalized 2-cochain v. Then $w = u - v$ is a 2-cocycle. As in Sect. 3.4 we have

$$0 = (\delta w)(a, 1, 1) = w(1, 1)^a - w(a1, 1) + w(a, 11) - w(a, 1)^1,$$

so that

$$w(a, 1) = w(1, 1)^a$$

for all $a \in S$. A 1-cochain $p \in C^1(S, \mathcal{G}'_\varphi)$ is now defined by

$$p(a) = w(a, 1)$$

for all $a \in S$. Then

$$(w - \delta p)(a, 1) = w(a, 1) - p(1)^a + p(a1) - p(a)^1 = w(a, 1) - w(1, 1)^a = 0$$

for all $a \in S$, so that $w - \delta p$ is normalized. Then $u - v - \delta p$ is normalized; since v is normalized, it follows that $u - \delta p$ is normalized. Moreover, $p(a) = w(a, 1) = u(a, 1) - v(a, 1) = u(a, 1)$, so that (5.30) yields $f^{\varphi(a)} = -u(a, 1) = -p(a)$ and $f + p(1) = 0$. Hence $(\varphi, \sigma, u - \delta p, 0)$ is a morphism from $(S, \mathcal{G}, t)$ to $(S', \mathcal{G}', t')$ which is normalized and isomorphic to Φ. $\square$

A morphism $\mathfrak{F} = (\mathcal{F}, \zeta, f)\colon \mathfrak{M} \longrightarrow \mathfrak{M}'$ in $\mathcal{R}$ is *normalized* if and only if $\zeta(X, I) = 1_X$ for all X; equivalently, in case $\mathfrak{M}'$ is reduced, if and only if $f = 1_I$, by (5.18). Hence $\mathfrak{F}$ is normalized if and only if $\mathbb{T}(\mathfrak{F})$ is normalized. Propositions 5.3.1 and 5.3.2 then yield

Proposition 5.3.3 *Every morphism of reduced commutative monoidal abelian groupoids is isomorphic to a normalized morphism.*

This also follows from Lemma 3.1 of [12].

5.3.2 Equivalence

Recall that two morphisms $\mathfrak{M}$ and $\mathfrak{M}'$ are *equivalent* in $\mathcal{R}$ if and only if there exist morphisms $\mathfrak{F}\colon \mathfrak{M} \longrightarrow \mathfrak{M}'$ and $\mathfrak{F}'\colon \mathfrak{M}' \longrightarrow \mathfrak{M}$ such that $\mathfrak{F}' \circ \mathfrak{F} \cong 1_{\mathfrak{M}}$ and $\mathfrak{F} \circ \mathfrak{F}' \cong 1_{\mathfrak{M}'}$.

Similarly, two symmetric 3-cocycle objects T and T' are *equivalent* in $\mathcal{T}$ if and only if there exist morphisms $\Phi\colon T \longrightarrow T'$ and $\Phi'\colon T' \longrightarrow T$ such that $\Phi' \circ \Phi \cong 1_T$ and $\Phi \circ \Phi' \cong 1_{T'}$.

Since $\mathbb{T}$ preserves composition and isomorphisms, we have

Proposition 5.3.4 *Two objects $\mathfrak{M}$ and $\mathfrak{M}'$ of $\mathcal{R}$ are equivalent in $\mathcal{R}$ if and only if $\mathbb{T}(\mathfrak{M})$ and $\mathbb{T}(\mathfrak{M}')$ are equivalent in $\mathcal{T}$.*

By Theorem 5.1.1, every commutative monoidal abelian groupoid is equivalent to a reduced commutative monoidal abelian groupoid. Hence Proposition 5.3.1, 5.3.4 and Theorem 5.2.1 yield:

Theorem 5.3.1 (First Classification Theorem) *There is a one-to-one correspondence between equivalence classes of symmetric 3-cocycle objects and equivalence classes of commutative monoidal abelian groupoids.*

5.3.3   Lone Cocycles

In the rest of this section, S is a given commutative monoid and $\mathcal{G}$ is a given abelian group valued functor on S. The Second Classification Theorem is obtained by replacing $\mathcal{R}$ and $\mathcal{T}$ by smaller subcategories.

Let $\mathcal{R}(S, \mathcal{G})$ be the subcategory of $\mathcal{R}$ whose objects are all objects of $\mathcal{R}$ with base $(S, \mathcal{G})$, in which a morphism $\mathfrak{F} = (\mathcal{F}, \zeta, f)$ from $\mathfrak{M} = (\mathcal{M}, \cdot, I, \mathfrak{a})$ to $\mathfrak{M}'$ in $\mathcal{R}(S, \mathcal{G})$ is a morphism in $\mathcal{R}$ such that $\mathcal{F}$ is the identity on $\mathcal{M}$.

Let $\mathcal{T}(S, \mathcal{G})$ be the category whose objects are all normalized 3-cocycles $t \in Z^3_{NS}(S, \mathcal{G})$, in which a morphism from t to t' is a pair (u, f) such that $u \in C^2_S(S, \mathcal{G})$, $f \in G_1$, and

$$t' = t - \delta u \quad \text{and} \quad u(a, 1) = -f^a \quad \text{for all } a \in S. \tag{5.35}$$

Composition in $\mathcal{T}(S, \mathcal{G})$ is

$$(v, g) \circ (u, f) = (u + v, \ f + g); \tag{5.36}$$

the identity morphism on t is $(0, 0)$.

The category $\mathcal{T}(S, \mathcal{G})$ comes with a functor into $\mathcal{T}$ that sends $t \in Z^3_{NS}(S, \mathcal{G})$ to its image $(S, \mathcal{G}, t)$ in $\mathcal{T}$, and $(u, f) \colon t \longrightarrow t'$ to its image $(1_S, 1_{\mathcal{G}}, u, f) \colon (S, \mathcal{G}, t) \longrightarrow (S, \mathcal{G}, t')$ in $\mathcal{T}$, which is a morphism since (5.29) and (5.30) then follow from (5.35). Composition and identity morphisms are preserved since (5.31) follows from (5.36). Thus, $\mathcal{T}(S, \mathcal{G})$ is isomorphic to a subcategory of $\mathcal{T}$.

A number of previous results yield easy corollaries. Theorem 5.2.1 yields:

Proposition 5.3.5 *The restriction of $\mathbb{T}$ to $\mathcal{R}(S, \mathcal{G})$ is an isomorphism of $\mathcal{R}(S, \mathcal{G})$ onto $\mathcal{T}(S, \mathcal{G})$.*

Two coterminal morphisms (u, f) and (u', f') from t to t' are *isomorphic* in $\mathcal{T}(S, \mathcal{G})$ if and only if their images in $\mathcal{T}$ are isomorphic; equivalently, if there exists a 1-cochain $p \in C^1(S, \mathcal{G}'_\varphi)$ such that (5.34) holds:

$$u' = u - \delta p \text{ and } f' = f + p(1).$$

A morphism (u, f) in $\mathcal{T}(S, \mathcal{G})$ is *normalized* if and only if its image in $\mathcal{T}$ is normalized; equivalently, if u is normalized $(u(a, 1) = 0$ for all $a \in S)$; equivalently, by (5.30), if $f = 0$. Proposition 5.3.2 yields:

Proposition 5.3.6 *Every morphism in $\mathcal{R}(S, \mathcal{G})$ is isomorphic to a normalized morphism.*

Two cocycles $t, t' \in Z_{NS}(S, \mathcal{G})$ are *equivalent* in $\mathcal{T}(S, \mathcal{G})$ if and only if their images in $\mathcal{T}$ are equivalent: if and only if there exist morphisms $(u, f) : t \longrightarrow t'$ and $(u', f') : t' \longrightarrow t$ such that $(u', f') \circ (u, f) \cong 1_t$ and $(u, f) \circ (u', f') \cong 1_{t'}$. By prop. 5.3.4,

Proposition 5.3.7 *Two objects $\mathfrak{M}$ and $\mathfrak{M}'$ of $\mathcal{R}(S, \mathcal{G})$ are equivalent in $\mathcal{R}(S, \mathcal{G})$ if and only if $\mathbb{T}(\mathfrak{M})$ and $\mathbb{T}(\mathfrak{M}')$ are equivalent in $\mathcal{T}(S, \mathcal{G})$.*

Unraveling the definitions we obtain:

Proposition 5.3.8 *Two cocycles $t, t' \in Z^3_{NS}(S, \mathcal{G})$ are equivalent in $\mathcal{T}(S, \mathcal{G})$ if and only if $t' - t$ is a 3-coboundary.*

Proof By definition, t and t' are equivalent if and only if there exist morphisms $\Phi : t \longrightarrow t'$ and $\Phi' : t' \longrightarrow t$ such that $\Phi' \circ \Phi$ is isomorphic to 1_t and $\Phi \circ \Phi'$ is isomorphic to $1_{t'}$. By Proposition 5.3.6, Φ and Φ' are isomorphic to normalized morphisms Ψ and Ψ'. Then it follows from Lemma 5.3.1 that $\Psi' \circ \Psi \cong 1_t$ and $\Psi \circ \Psi' \cong 1_{t'}$. In other words, we may assume that Φ and Φ' are normalized. Then $\Phi = (u, 0)$ and $\Phi' = (u', 0)$, where u and u' are normalized, and (5.35) yields $t' = t - \delta u$.

Conversely assume that $t' = t - \delta u$ for some 2-cochain $u \in C^2_S(S, \mathcal{G})$. Then δu is normalized; by Lemma 3.4.4 we may assume that u is normalized. Then $(u, 0)$ is a morphism from t to t' and $(-u, 0)$ is a morphism from t' to t. Moreover, $(-u, 0) \circ (u, 0) = (0, 0) = 1_t$ and $(u, 0) \circ (-u, 0) = (0, 0) = 1_{t'}$. Therefore t and t' are equivalent. $\square$

Propositions 5.3.7 and 5.3.8 yield:

Theorem 5.3.2 (Second Classification Theorem) *For every commutative monoid S and abelian group valued functor $\mathcal{G}$ on S, there is a one-to-one correspondence between the elements of $H^3_S(S, \mathcal{G})$ and the equivalence classes of commutative monoidal abelian groupoids with base $(S, \mathcal{G})$.*

5.4   Braided Groupoids

Braided monoidal abelian groupoids are somewhat more general than commutative monoidal abelian groupoids and are classified by the larger cohomology group $H^3_C(S, \mathcal{G})$.

5.4.1   Definition

A braided monoidal abelian groupoid is a braided monoidal category (as in App. 9.4) whose underlying category $\mathcal{M}$ is an abelian groupoid. With weaker commutativity the natural isomorphism $\mathfrak{c}$ is now included in the definition, which follows Joyal and Street [36]; Calvo-Cervera and Cegarra's definition in [10] is equivalent but slightly different.

Thus a *braided monoidal abelian groupoid*

$$\mathfrak{M} = (\mathcal{M}, \cdot, I, \mathfrak{a}, \mathfrak{l}, \mathfrak{r}, \mathfrak{c})$$

(called a *braided abelian $\otimes$-groupoid* in [10]) consists of an abelian groupoid $\mathcal{M}$, a multiplication $\cdot$ on $\mathcal{M}$, an identity object I, and natural isomorphisms

$$\mathfrak{a} = \mathfrak{a}(X, Y, Z) \colon (XY)Z \longrightarrow X(YZ),$$

$$\mathfrak{l} = \mathfrak{l}(X) \colon IX \longrightarrow X,$$

$$\mathfrak{r} = \mathfrak{r}(X) \colon XI \longrightarrow X, \quad \text{and}$$

$$\mathfrak{c} = \mathfrak{c}(X, Y) \colon XY \longrightarrow YX,$$

with property (9.44), which is the same as (5.2):

$$\mathfrak{a}(X, Y, ZT) \circ \mathfrak{a}(XY, Z, T) \tag{5.37}$$
$$= 1_X \cdot \mathfrak{a}(Y, Z, T) \circ \mathfrak{a}(X, YZ, T) \circ \mathfrak{a}(X, Y, Z) \cdot 1_T;$$

property (9.45), which is the same as (5.3):

$$1_X \cdot \mathfrak{r}(Y) \circ \mathfrak{a}(X, I, Y) = \mathfrak{r}(X) \cdot 1_Y; \tag{5.38}$$

property (9.46):

$$\mathfrak{l}(I) = \mathfrak{r}(I) \colon II \longrightarrow I; \tag{5.39}$$

property (9.53), which is the same as (5.4):

$$\mathfrak{a}(Y, Z, X) \circ \mathfrak{c}(X, YZ) \circ \mathfrak{a}(X, Y, Z) \qquad (5.40)$$
$$= 1_Y \cdot \mathfrak{c}(X, Z) \circ \mathfrak{a}(Y, X, Z) \circ \mathfrak{c}(X, Y) \cdot 1_Z ;$$

and property (9.54), which is the same as (5.5):

$$\mathfrak{a}(Z, X, Y)^{-1} \circ \mathfrak{c}(XY, Z) \circ \mathfrak{a}(X, Y, Z)^{-1} \qquad (5.41)$$
$$= \mathfrak{c}(X, Z) \cdot 1_Y \circ \mathfrak{a}(X, Z, Y)^{-1} \circ 1_X \cdot \mathfrak{c}(Y, Z) .$$

These properties imply:

$$\mathfrak{l}(X) \circ \mathfrak{c}(X, I) = \mathfrak{r}(X) \text{ and} \qquad (5.42)$$
$$\mathfrak{r}(X) \circ \mathfrak{c}(I, X) = \mathfrak{l}(X) . \qquad (5.43)$$

As in App. 9.4, these properties also imply (9.47) and (9.48):

$$\mathfrak{l}(YZ) \circ \mathfrak{a}(I, Y, Z) = \mathfrak{l}(Y) \cdot 1_Z , \qquad (5.44)$$
$$(1_X \cdot \mathfrak{r}(Y)) \circ \mathfrak{a}(X, Y, I) = \mathfrak{r}(XY) . \qquad (5.45)$$

A *morphism* $\mathfrak{F} = (\mathcal{F}, \zeta, f)$ of braided monoidal abelian groupoids from $\mathfrak{M} = (\mathcal{M}, \cdot, I, \mathfrak{a}, \mathfrak{l}, \mathfrak{r}, \mathfrak{c})$ to $\mathfrak{M}' = (\mathcal{M}', \cdot, I', \mathfrak{a}', \mathfrak{l}', \mathfrak{r}', \mathfrak{c}')$ (called a *braided monoidal functor* in [42] and App. 9.4, a *braided $\otimes$-functor* in [10]) consists of a functor $\mathcal{F}\colon \mathcal{M} \longrightarrow \mathcal{M}'$, a natural isomorphism $\zeta(X, Y)\colon \mathcal{F}X \cdot \mathcal{F}Y \longrightarrow \mathcal{F}(XY)$, and an isomorphism $f\colon I' \longrightarrow \mathcal{F}I$, with property (9.49), which is the same as (5.9):

$$\mathcal{F}\mathfrak{a}(X, Y, Z) \circ \zeta(XY, Z) \circ (\zeta(X, Y) \cdot 1_{\mathcal{F}Z}) \qquad (5.46)$$
$$= \zeta(X, YZ) \circ (1_{\mathcal{F}X} \cdot \zeta(Y, Z)) \circ \mathfrak{a}'(\mathcal{F}X, \mathcal{F}Y, \mathcal{F}Z) ,$$

property (9.50), which includes property (5.10):

$$\mathcal{F}\mathfrak{l}(X) \circ \zeta(I, X) \circ (f \cdot 1_{\mathcal{F}X}) = \mathfrak{l}'(\mathcal{F}X) \text{ and} \qquad (5.47)$$
$$\mathcal{F}\mathfrak{r}(X) \circ \zeta(X, I) \circ (1_{\mathcal{F}X} \cdot f) = \mathfrak{r}'(\mathcal{F}X) , \qquad (5.48)$$

and property (9.56):

$$\mathcal{F}\mathfrak{c}(X, Y) \circ \zeta(X, Y) = \zeta(Y, X) \circ \mathfrak{c}'(\mathcal{F}X, \mathcal{F}Y) , \qquad (5.49)$$

for all objects X, Y, Z of $\mathcal{M}$.

Morphisms of braided monoidal abelian groupoids compose according to (5.12): if $\mathfrak{F} = (\mathcal{F}, \zeta, f)\colon \mathfrak{M} \longrightarrow \mathfrak{M}'$ and $\mathfrak{F}' = (\mathcal{F}', \zeta', f')\colon \mathfrak{M}' \longrightarrow \mathfrak{M}''$ are morphisms, then $\mathfrak{F}' \circ \mathfrak{F} = (\mathcal{F}'', \zeta'', f'')$, where:

$$\mathcal{F}'' = \mathcal{F}' \circ \mathcal{F}, \tag{5.50}$$

$$\zeta''(X, Y) = \mathcal{F}'\zeta(\mathcal{F}X, \mathcal{F}Y) + \zeta'(\mathcal{F}X, \mathcal{F}Y), \quad \text{and}$$

$$f'' = \mathcal{F}'f + f',$$

is a morphism $\mathfrak{F}'' \colon \mathfrak{M} \longrightarrow \mathfrak{M}''$. The identity morphism on $(\mathcal{M}, \cdot, I, \mathfrak{a}, \mathfrak{l}, \mathfrak{r}, \mathfrak{c})$ is $(\mathcal{F}, \zeta, f)$, where $\mathcal{F}$ is the identity on $\mathcal{M}$, $\zeta(X, Y) = 1_{XY}$, and $f = 1_I$.

Braided monoidal abelian groupoids and their morphisms now constitute a category $\mathcal{B}$.

5.4.2 Reduction

As in Sect. 5.1, two morphisms $\mathfrak{F} = (\mathcal{F}, \zeta, f)$ and $\mathfrak{F}' = (\mathcal{F}', \zeta', f')$ of $\mathcal{B}$ from $\mathfrak{M} = (\mathcal{M}, \cdot, I, \mathfrak{a}, \mathfrak{l}, \mathfrak{r}, \mathfrak{c})$ to $\mathfrak{M}' = (\mathcal{M}', \cdot, I', \mathfrak{a}', \mathfrak{l}', \mathfrak{r}', \mathfrak{c}')$ are *isomorphic* if and only if there exists a natural isomorphism $\theta \colon \mathcal{F} \longrightarrow \mathcal{F}'$ (called a *braided isomorphism* in [10]) such that

$$\theta(XY) \circ \zeta(X, Y) = \zeta'(X, Y) \circ \theta(X) \cdot \theta(Y) \quad \text{and} \tag{5.51}$$

$$\theta(I) \circ f = f'$$

for all objects X, Y of $\mathcal{M}$. Two braided monoidal abelian groupoids $\mathfrak{M}$ and $\mathfrak{M}'$ are *equivalent* if and only if there exist morphisms $\mathfrak{F} \colon \mathfrak{M} \longrightarrow \mathfrak{M}'$ and $\mathfrak{F}' \colon \mathfrak{M}' \longrightarrow \mathfrak{M}$ such that $\mathfrak{F}' \circ \mathfrak{F} \cong 1_{\mathfrak{M}}$ and $\mathfrak{F} \circ \mathfrak{F}' \cong 1_{\mathfrak{M}'}$.

We call a braided monoidal abelian groupoid $\mathfrak{M} = (\mathcal{M}, \cdot, I, \mathfrak{a}, \mathfrak{l}, \mathfrak{r}, \mathfrak{c})$ *reduced* if and only if $\mathcal{M}$ is totally disconnected and $\mathfrak{l}, \mathfrak{r}$ are both identity morphisms ($\mathfrak{l}(X) = 1_X$ and $\mathfrak{r}(X) = 1_X$). Reduced braided monoidal abelian groupoids are called *totally disconnected and strictly unitary* in [10].

The first main result in this section is:

Theorem 5.4.1 *Every braided monoidal abelian groupoid is equivalent to a reduced braided monoidal abelian groupoid.*

The proof is outlined in [10] on p.1024.

If $\mathfrak{M} = (\mathcal{M}, \cdot, I, \mathfrak{a}, \mathfrak{l}, \mathfrak{r}, \mathfrak{c})$ is a reduced braided monoidal abelian groupoid, then the isomorphisms $\mathfrak{l}$ and $\mathfrak{r}$ do not need to be specified, and we just write $\mathfrak{M} = (\mathcal{M}, \cdot, I, \mathfrak{a}, \mathfrak{c})$; moreover, conditions (5.38), (5.44), (5.45), and (5.42) become simpler:

$$\mathfrak{a}(X, I, Y) = 1_{XY}, \tag{5.52}$$

$$\mathfrak{a}(I, Y, Z) = 1_{YZ}, \tag{5.53}$$

$$a(X, Y, I) = 1_{XY}, \tag{5.54}$$

$$c(X, I) = 1_X, \tag{5.55}$$

whereas (5.37), (5.41), and (5.40) remain unchanged.

Similarly, conditions (5.46) and (5.49) for a morphism $\mathfrak{M} \longrightarrow \mathfrak{M}'$ remain unchanged when $\mathfrak{M}$ and $\mathfrak{M}'$ are reduced, and conditions (5.47) and (5.48) simplify to:

$$\zeta(I, X) \circ (f \cdot 1_{\mathcal{F}X}) = 1_{\mathcal{F}X} \text{ and} \tag{5.56}$$

$$\zeta(X, I) \circ (1_{\mathcal{F}X} \cdot f) = 1_{\mathcal{F}X}. \tag{5.57}$$

5.4.3 The Base

If $\mathfrak{M} = (\mathcal{M}, \cdot, I, a, c)$ is a reduced braided monoidal abelian groupoid, then, as in Sect. 5.1, the set S of objects of $\mathcal{M}$ is a commutative monoid under $\cdot$ with identity element I, and $\mathcal{G} = (G, \gamma)$, where $G_X = \mathrm{Aut}_{\mathcal{M}}(X)$ and $\gamma_{X,Y} \colon G_X \longrightarrow G_{XY}$, $g \longmapsto g^Y = g \cdot 1_Y$, is an abelian group valued functor on S. Furthermore, $g \cdot h = g^Y \circ h^X$ for all $g \in G_X$ and $h \in G_Y$, so that $\mathcal{M}$, $\cdot$, and I are completely determined by S and $\mathcal{G}$. This is proved like Prop. 1.1.

The pair $(S, \mathcal{G})$ is the *base* of $\mathfrak{M}$.

At this point we start writing the abelian groups G_X additively.

Lemma 5.4.1 *If $\mathfrak{M} = (\mathcal{M}, \cdot, I, a, c)$ is a reduced braided monoidal abelian groupoid with base $(S, \mathcal{G})$, then $(a, c) \in Z_e^3(S, \mathcal{G})$ is an extended 3-cocycle.*

Proof First, $a \in C^3(S, \mathcal{G})$, since $a(X, Y, Z) \in G_{XYZ}$ for all X, Y, Z, and it follows from (5.53) (or from (5.54)) that a is normalized. Similarly, $c \in C^2(S, \mathcal{G})$, since $c(X, Y) \in G_{XY}$ for all X, Y, and it follows from (5.55) that c is normalized. Hence $(a, c) \in C_e^3(S, \mathcal{G}) = C_N^3(S, \mathcal{G}) \times C_N^2(S, \mathcal{G})$.

Recall that the coboundary of an extended 3-cochain (t, u) is the extended 4-cochain $\delta(t, u) = (q, r, s)$, where

$$q(a, b, c, d) = t(b, c, d)^a - t(ab, c, d)$$

$$+ t(a, bc, d) - t(a, b, cd) + t(a, b, c)^d = (\delta t)(a, b, c, d),$$

$$r(a, b, c) = u(a, c)^b - u(a, bc) + u(a, b)^c$$

$$- t(a, b, c) + t(b, a, c) - t(b, c, a),$$

$$s(a, b, c) = u(b, c)^a - u(ab, c) + u(a, c)^b$$

$$+ t(a, b, c) - t(a, c, b) + t(c, a, b),$$

for all $a, b, c, d \in S$. If $(t, u) = (\mathfrak{a}, \mathfrak{c})$, then $q = \delta\mathfrak{a} = 0$, by (5.37);

$$
\begin{aligned}
r(X, Y, Z) \;=\; & \mathfrak{c}(X, Z)^Y - \mathfrak{c}(X, YZ) + \mathfrak{c}(X, Y)^Z \\
& - \mathfrak{a}(X, Y, Z) + \mathfrak{a}(Y, X, Z) - \mathfrak{a}(Y, Z, X) \;=\; 0 \,,
\end{aligned}
$$

by (5.40); and

$$
\begin{aligned}
s(X, Y, Z) \;=\; & \mathfrak{c}(Y, Z)^X - \mathfrak{c}(XY, Z) + \mathfrak{c}(X, Z)^Y \\
& + \mathfrak{a}(X, Y, Z) - \mathfrak{a}(X, Z, Y) + \mathfrak{a}(Z, X, Y) \;=\; 0 \,,
\end{aligned}
$$

by (5.41). Thus $\delta(\mathfrak{a}, \mathfrak{c}) = 0$.   □

5.4.4 Extended Cocycle Objects

As in Sect. 5.3 we bundle the base of $\mathfrak{M}$ and its extended 3-cocycle into an *extended 3-cocycle object* $(S, \mathcal{G}, e)$ that consists of a commutative monoid S, an abelian group valued functor $\mathcal{G}$ on S, and an extended 3-cocycle $e \in Z_e^3(S, \mathcal{G})$ on S with values in $\mathcal{G}$. Lemma 5.4.1 can now be restated as the direct part of:

Proposition 5.4.1 *If* $\mathfrak{M} = (\mathcal{M}, \cdot, I, \mathfrak{a}, \mathfrak{c})$ *is a reduced braided monoidal abelian groupoid with base* $(S, \mathcal{G})$, *then*

$$
\mathbb{E}(\mathfrak{M}) \;=\; (S, \mathcal{G}, (\mathfrak{a}, \mathfrak{c}))
$$

is an extended 3-cocycle object. Conversely, for every extended 3-cocycle object E, *there exists a reduced braided monoidal abelian groupoid* $\mathfrak{M}$ *such that* $\mathbb{E}(\mathfrak{M}) = $ E; *and* $\mathfrak{M}$ *is unique.*

Extended 3-cocycle objects are themselves the objects of a category $\mathcal{E}$, in which a morphism (φ, σ, r, f) from E $= (S, \mathcal{G}, (t, u))$ to E$' = (S', \mathcal{G}', (t', u'))$ consists of a homomorphism $\varphi \colon S \longrightarrow S'$, a natural transformation $\sigma \colon \mathcal{G} \longrightarrow \mathcal{G}'_\varphi$, a 2-cochain $r \in C^2(S, \mathcal{G}'_\varphi)$, and $f \in G'_1$, such that

$$
\varphi^* t' \;=\; \sigma_* t - \delta r, \tag{5.58}
$$

$$
r(1, a) \;=\; r(a, 1) \;=\; - f^{\varphi(a)}, \text{ and} \tag{5.59}
$$

$$
(\varphi^* u')(a, b) \;=\; (\sigma_* u)(a, b) + r(a, b) - r(b, a) \,, \tag{5.60}
$$

for all $a, b \in S$, where $\mathcal{G}'_\varphi$, φ^*, and σ_* were defined in Sect. 5.2. Composition follows (5.31):

$$
(\psi, \tau, v, g) \circ (\varphi, \sigma, u, f) \;=\; (\psi \circ \varphi, \; \tau^\varphi \circ \sigma, \; \tau_*^\varphi u + \varphi^* v, \; \tau(1) f + g). \tag{5.61}
$$

The identity morphism on $E = (S, \mathcal{G}, e)$ is $(1_S, 1_\mathcal{G}, 0, 0)$. Extended 3-cocycle objects and their morphisms now constitute a category $\mathcal{E}$.

If $(\mathcal{F}, \zeta, f)$ is a morphism from $\mathfrak{M}$ to $\mathfrak{M}'$, where $\mathfrak{M}$ and $\mathfrak{M}'$ are reduced, then $\mathbb{E}(\mathcal{F}, \zeta, f) = (\varphi, \sigma, \zeta, f)$ is a morphism from $\mathbb{E}(\mathfrak{M})$ to $\mathbb{E}(\mathfrak{M}')$, in which $\varphi(X) = \mathcal{F}X$ and $\sigma(X)g = \mathcal{F}g$ for all X and g; the conditions above follow from (5.46), (5.56), (5.57), and (5.49). This makes $\mathbb{E}$ a functor. In fact, $\mathbb{E}$ is an isomorphism of categories (this is implicit in [10]). Thus:

Theorem 5.4.2 *The category $\mathcal{B}$ of reduced braided monoidal abelian groupoids is isomorphic to the category $\mathcal{E}$ of extended 3-cocycle objects.*

5.4.5 Classification

Two (coterminal) morphisms $\Phi = (\varphi, \sigma, r, f)$ and $\Phi' = (\varphi', \sigma', r', f')$ of $\mathcal{E}$ are *isomorphic* if $\varphi' = \varphi, \sigma' = \sigma$, and there exists a 1-cochain $p \in C^1(S, \mathcal{G}'_\varphi)$ such that

$$r' = r - \delta p \quad \text{and} \quad f' = f + p(1). \tag{5.62}$$

Condition (5.62) matches (5.51), so that two coterminal morphisms $\mathfrak{F}$ and $\mathfrak{F}'$ of $\mathcal{B}$ are isomorphic if and only if $\mathbb{E}(\mathfrak{F})$ and $\mathbb{E}(\mathfrak{F}')$ are isomorphic in $\mathcal{E}$.

Two extended 3-cocycle objects $E = (S, \mathcal{G}, e)$ and $E' = (S', \mathcal{G}', e')$ are *equivalent* if and only if there exist morphisms $\Phi: E \longrightarrow E'$ and $\Phi': E' \longrightarrow E$ such that $\Phi' \circ \Phi \cong 1_E$ and $\Phi \circ \Phi' \cong 1_{E'}$. Since $\mathbb{E}$ preserves composition and isomorphism, two reduced braided monoidal abelian groupoids $\mathfrak{M}$ and $\mathfrak{M}'$ are equivalent if and only if the corresponding extended 3-cocycle objects $\mathbb{E}(\mathfrak{M})$ and $\mathbb{E}(\mathfrak{M}')$ are equivalent. Hence Theorem 5.4.1 yields

Theorem 5.4.3 (First Classification Theorem, Extended Version) *There is a one-to-one correspondence between equivalence classes of braided monoidal abelian groupoids and equivalence classes of extended 3-cocycle objects.*

For the Second Classification Theorem we let S be a given commutative monoid and let $\mathcal{G}$ be a given abelian group valued functor on S.

Let $\mathcal{B}(S, \mathcal{G})$ be the subcategory of $\mathcal{B}$ whose objects are the objects of $\mathcal{B}$ with base $(S, \mathcal{G})$, in which a morphism $\mathfrak{F} = (\mathcal{F}, \zeta, f)$ from $\mathfrak{M} = (\mathcal{M}, \cdot, I, \mathfrak{a}, \mathfrak{c})$ to $\mathfrak{M}'$ is a morphism of $\mathcal{B}$ such that $\mathcal{F}$ is the identity on $\mathcal{M}$.

Extended 3-cocycles are the objects of a category $\mathcal{E}(S, \mathcal{G})$, in which a morphism from (t, u) to (t', u') is a pair (r, f) such that $u \in C_S^2(S, \mathcal{G})$, $f \in G_1$, and (5.58), (5.59), and (5.60) hold:

$$t' = t - \delta r \tag{5.63}$$

$$r(1, a) = r(a, 1) = -f^a, \quad \text{and} \tag{5.64}$$

$$u'(a, b) = u(a, b) + r(a, b) - r(b, a), \tag{5.65}$$

for all $a, b \in S$. Morphisms of $\mathcal{E}(S, \mathcal{G})$ compose by:

$$(r, f) \circ (s, g) = (r + s, f + g). \tag{5.66}$$

The identity morphism on (t, u) is $(0, 0)$.

The category $\mathcal{E}(S, \mathcal{G})$ comes with a functor into $\mathcal{E}$ that sends $e \in Z_e^3(S, \mathcal{G})$ to its image $(S, \mathcal{G}, e)$ in $\mathcal{E}$, and $(r, f): e \longrightarrow e'$ to its image $(1_S, 1_\mathcal{G}, r, f): (S, \mathcal{G}, e) \longrightarrow (S, \mathcal{G}, e')$ in $\mathcal{E}$, which is a morphism since (5.58), (5.59), and (5.60) then follow from (5.63), (5.64), and (5.65). Composition and identity morphisms are preserved since (5.61) follows from (5.66). Thus, $\mathcal{E}(S, \mathcal{G})$ is isomorphic to a subcategory of $\mathcal{E}$ and is therefore isomorphic to the subcategory $\mathcal{B}(S, \mathcal{G})$ of $\mathcal{B}$.

Two coterminal morphisms (r, f) and (r', f') are *isomorphic* in $\mathcal{E}(S, \mathcal{G})$ if and only if they are isomorphic in $\mathcal{E}$; equivalently, if there exists a 1-cochain $p \in C^1(S, \mathcal{G})$ such that (5.62) holds:

$$r' = r - \delta p \text{ and } f' = f + p(1).$$

A morphism (r, f) in $\mathcal{E}(S, \mathcal{G})$ is *normalized* if and only if r is normalized: $r(a, 1) = r(1, a) = 0$ for all $a \in S$; equivalently, by (5.59), if $f = 0$.

Proposition 5.4.2 *Every morphism in $\mathcal{E}(S, \mathcal{G})$ is isomorphic to a normalized morphism.*

This also follows from [12].

Proof Let $(r, f): (t, u) \longrightarrow (t', u')$. Then $\delta r = t - t'$ is normalized. Hence

$$0 = (\delta r)(a, 1, 1) = r(1, 1)^a - r(a, 11) + r(a1, 1) - r(a, 1)^1$$

and $r(a, 1) = r(1, 1)^a$ for all $a \in S$. Let $p(a) = r(a, 1)$. Then

$$(r - \delta p)(a, 1) = r(a, 1) - p(1)^a + p(a1) - p(a)^1 = 0,$$
$$(r - \delta p)(1, a) = r(1, a) - p(a)^1 + p(1a) - p(1)^a = 0, \text{ and}$$
$$f + p(1) = f + r(1, 1) = 0,$$

for all $a \in S$, by (5.64), so that $(r - \delta p, 0)$ is normalized and equivalent to (r, f).

$\square$

Since composition of morphisms preserves isomorphism it follows from Proposition 5.4.2 that two extended 3-cocycles e and e' are equivalent in $\mathcal{E}(S, \mathcal{G})$ if and only if there exist normalized morphisms $\Phi: e \longrightarrow e'$ and $\Phi': e' \longrightarrow e$ such that $\Phi' \circ \Phi \cong 1_e$ and $\Phi \circ \Phi' \cong 1_{e'}$.

Proposition 5.4.3 *Two extended 3-cocycles are equivalent in $\mathcal{E}(S, \mathcal{G})$ if and only if they differ by an extended coboundary.*

Proof Recall that the extended coboundary of a normalized 2-cochain $r \in C_e^2(S, \mathcal{G})$ is $\delta^e r = (\delta r, u)$, where $u(a, b) = r(b, a) - r(a, b)$.

If (t, u) and (t', u') are equivalent, then there exists a normalized morphism $(r, f): (t, u) \longrightarrow (t', u')$ and (5.63), (5.65) show that $(t, u) - (t', u') = \delta^e r$.

Conversely, if $(t, u) - (t', u') = \delta^e r$ for some normalized 2-cochain $r \in C_e^2(S, \mathcal{G})$, then $(r, 0)$ is a morphism from (t, u) to (t', u'), $(-r, 0)$ is a morphism from (t', u') to (t, u), $(-r, 0) \circ (r, 0)$ is the identity on (t, u), and $(r, 0) \circ (-r, 0)$ is the identity on (t', u'). Hence (t, u) and (t', u') are equivalent. □

Since $\mathcal{B}(S, \mathcal{G})$ and $\mathcal{E}(S, \mathcal{G})$ are isomorphic, Proposition 5.4.3 yields:

Theorem 5.4.4 (Second Classification Theorem, Extended Version) *For every commutative monoid S and abelian group valued functor $\mathcal{G}$ on S there is a one-to-one correspondence between the elements of $H_C^3(S, \mathcal{G})$ and the equivalence classes of braided monoidal abelian groupoids with base $(S, \mathcal{G})$.*

Chapter 6
The Overpath Method

The Overpath Method, developed in [22] and [26] (see also [31]), is a more efficient method that calculates $H^2(S, \mathcal{G})$ from a suitable presentation of S, which is most useful when S is finite or finitely generated.

6.1 Paths and Overpaths

The Overpath Method relies on the analysis of congruences on free commutative semigroups in [24] or [50], also found in [31] or [51], related to the analysis that leads to Gröbner bases (see e.g. [1, 7]).

6.1.1 Free Commutative Monoids

In this chapter, $\mathbb{F}$ is the free commutative monoid on some set $X \subseteq \mathbb{F}$ (a *basis* of $\mathbb{F}$). We write $\mathbb{F}$ additively, so that the typical element of $\mathbb{F}$ is uniquely a linear combination $a = \sum_{x \in X} a_x \, x$ with nonnegative integer coefficients $a_x \geq 0$ such that $a_x = 0$ for all but finitely $x \in X$.

Equivalently, $\mathbb{F}$ is the positive [or zero] cone of the free abelian group $\mathbb{G}$ on X, whose typical element is uniquely a linear combination $\sum_{x \in X} a_x \, x$ with (arbitrary) integer coefficients a_x such that $a_x = 0$ for all but finitely $x \in X$.

The *length* of $a \in \mathbb{F}$ (sometimes called the *degree* of a) is $|a| = \sum_{x \in X} a_x$ (this would be the length of a as a commutative word, were $\mathbb{F}$ written multiplicatively).

The semigroup $\mathbb{F}$ has a natural partial order $\leq$, under which $a \leq b$ if and only if $a_x \leq b_x$ for all $x \in X$; equivalently, $b = a + c$ for some $c \in \mathbb{F}$. Thus, $\leq$ is the opposite of the divisibility preorder on $\mathbb{F}$.

P. A. Grillet, *The Cohomology of Commutative Semigroups*, Lecture Notes in Mathematics 2307, https://doi.org/10.1007/978-3-031-08212-2_6

The following property is named after Dickson who proved it for the multiplicative subsemigroup of $\mathbb{N}$ generated by finitely many primes [14]. An *antichain* of $\mathbb{F}$ is a subset A of $\mathbb{F}$ such that $a \leq b$ implies $a = b$ if $a, b \in A$.

Proposition 6.1.1 (Dickson's Theorem) *If X is finite, then every antichain of $\mathbb{F}$ is finite.*   □

For a proof, see [13] or [31].

A *compatible well order* on $\mathbb{F}$ (also called an *admissible ordering*) is a total order relation $\prec$ such that: $(\mathbb{F}, \prec)$ is well-ordered; $a < b$ implies $a \prec b$; and $a \prec b$ implies $a + c \prec b + c$ for all $c \in \mathbb{F}$.

Every free commutative monoid has a compatible well order; in fact, compatible well orders on $\mathbb{F}$ can be constructed in various ways, especially if $\mathbb{F}$ is finitely generated [1, 7]. For example, if X is finite, then every total order on X induces two compatible well orders on $\mathbb{F}$:

the *lexicographic order*, under which $a \prec b$ if and only if $a \neq b$ and $a_y < b_y$, where y is the least element x of X such that $a_x \neq b_x$; and

the *degree lexicographic order*, under which $a \prec b$ if and only if either $|a| < |b|$, or $|a| = |b|$ and $a \prec b$ in the lexicographic order.

In either case, $x > y$ in X implies $x \prec y$ in $\mathbb{F}$.

6.1.2   Congruences

A compatible well order $\prec$ on $\mathbb{F}$ induces for every congruence $\mathcal{C}$ on $\mathbb{F}$ a cross-section of $\mathcal{C}$ and 'canonical' generators. For each $a \in \mathbb{F}$ let r_a be the least element (under $\prec$) of the $\mathcal{C}$-class C_a of a (this is the *function minimum* of Rosales [50]). Then

$$R = \{r_a \mid a \in \mathbb{F}\}$$

is a cross section of $\mathcal{C}$. Let

$$M = \{m \in \mathbb{F} \mid m \text{ is minimal (under } \leq \text{) such that } r_m \neq m\}.$$

Proposition 6.1.2 $\mathcal{C}$ *is generated by all pairs* (m, r_m) *with* $m \in M$.

This is proved in [24, 31, 50].

Proposition 6.1.2 presents $S \cong \mathbb{F}/\mathcal{C}$ as the commutative semigroup generated by the basis X of $\mathbb{F}$, subject to all relations (m, r_m). Since M is an antichain of $\mathbb{F}$, Propositions 6.1.1 and 6.1.2 imply *Rédei's Theorem*: if $\mathbb{F}$ is finitely generated, then every congruence on $\mathbb{F}$ is finitely generated. This was proved by Rédei (the hard way) in 1956 [48].

Each $m \in M$ comes with a *defining vector*

$$\overrightarrow{m} = m - r_m \in \mathbb{G}.$$

The defining vectors of $\mathcal{C}$ are related to its *Rédei group*

$$R = \{a - b \in \mathbb{G} \mid a \, \mathcal{C} \, b\}.$$

Proposition 6.1.3 *Relative to a compatible well order on* $\mathbb{F}$, *the subgroup of* $\mathbb{G}$ *generated by the defining vectors of* $\mathcal{C}$ *is the Rédei group* R *of* $\mathcal{C}$. *The universal group of* $S = \mathbb{F}/\mathcal{C}$ *is isomorphic to* $\mathbb{G}/R$.

Proof Let K be the subgroup of $\mathbb{G}$ generated by the defining vectors. If $a \, \mathcal{C} \, b$, then by Proposition 6.1.2 $a - b$ is a sum of differences $m - r_m$ and $r_m - m$, so that $a - b \in K$. Thus $R \subseteq K$. Conversely, R contains every defining vector $\overrightarrow{m}$, since $m \, \mathcal{C} \, r_m$. Hence $R = K$.

As a commutative monoid, S is generated by X subject to all defining relations $m = r_m$. Therefore the universal group $G(S)$ of S is the abelian group generated by X subject to the same defining relations $m = r_m$. Hence $G(S) \cong \mathbb{G}/K$.   □

The well ordering of $\mathbb{F}$ to select 'minimal' generators of $\mathcal{C}$ is reminiscent of Gröbner bases. There is a more direct link. Let K be a field and let $K[X]$ be the polynomial ring with the set X of commuting indeterminates. Ordering $\mathbb{F}$ also orders the monomials

$$X^a = \prod_{x \in X} x^{a_x} \in K[X], \quad \text{where} \quad a = \sum_{x \in X} a_x \, x \in \mathbb{F}.$$

The congruence $\mathcal{C}$ on $\mathbb{F}$ induces an ideal $I(\mathcal{C})$ of $K[X]$, which is generated by all $X^a - X^b$ such that $a \, \mathcal{C} \, b$. (In turn, $I(\mathcal{C})$ induces $\mathcal{C}$ on $\mathbb{F}$ [44].) In the above, the set $\{X^m - X^{r_m} \mid m \in M\}$ is a Gröbner basis of $I(\mathcal{C})$ (Prop. XII.1.6 of [31].)

6.1.3   Paths

The elements of $\mathbb{F}$ are the nodes of a directed graph with labeled edges, in which there is an *edge* $a \xrightarrow{m} b$ labeled m if and only if $m \in M$ and there exists $t \in \mathbb{F}$ such that $a = m + t$ and $b = r_m + t$; then $a \geq m, b \prec a, a \, \mathcal{C} \, b$, and $a - b = m - r_m = \overrightarrow{m} \in \mathbb{G}$, so that $b = a - \overrightarrow{m}$.

Relative to a compatible well order on $\mathbb{F}$, a descending *path* $\mathfrak{p} \colon a \longrightarrow b$ from $a \in \mathbb{F}$ to $b \in \mathbb{F}$ of length $n \geq 0$ is a sequence of edges

$$\mathfrak{p} \colon a = a_0 \xrightarrow{m_1} a_1 \xrightarrow{m_2} a_2 \longrightarrow \cdots \longrightarrow a_{n-1} \xrightarrow{m_n} a_n = b,$$

where $a_1, \ldots, a_n \in \mathbb{F}$ and $m_1, \ldots, m_n \in M$. If $n = 0$, then $\mathfrak{p} \colon a \longrightarrow a$ is the *empty path* from a to a. If $n = 1$, then $\mathfrak{p}$ is an edge. If $n = 1$ and $a = m \in M$, then $\mathfrak{p}$ is the *simple path*

$$\mathfrak{p} \colon m \xrightarrow{m} r_m.$$

If there is a path from a to b, then $a \mathrel{\mathcal{C}} b$; if $n = 0$, then $a = b$; if $n > 0$, then $a \succ b$, all intermediate nodes $a_1, \ldots, a_n$ lie in the $\mathcal{C}$-class of a, and $a \succ a_1 \succ a_2 \succ \cdots \succ a_n = b$.

Proposition 6.1.4 *For each $a \in \mathbb{F}$ there exists a path from a to r_a.*

This is readily proved by ordinal induction, using Proposition 6.1.2 [22, 31]. There are two basic operations on paths. Paths can be added: if

$$\mathfrak{p} \colon a = a_0 \xrightarrow{m_1} a_1 \longrightarrow \cdots \xrightarrow{m_k} a_k = b$$

is a path from a to b, and

$$\mathfrak{q} \colon b = b_0 \xrightarrow{n_1} b_1 \longrightarrow \cdots \xrightarrow{n_\ell} b_\ell = c$$

is a path from b to c, then

$$\mathfrak{p} + \mathfrak{q} \colon a \xrightarrow{m_1} a_1 \longrightarrow \cdots \xrightarrow{m_k} a_k = b \xrightarrow{n_1} b_1 \longrightarrow \cdots \xrightarrow{n_\ell} b_\ell = c$$

is a path from b to c. In effect this turns the elements of $\mathbb{F}$ into the objects of a category, whose morphisms are paths.

Paths can also be translated: if $t \in \mathbb{F}$ and

$$\mathfrak{p} \colon a = a_0 \xrightarrow{m_1} a_1 \longrightarrow \cdots \xrightarrow{m_k} a_k = b$$

is a path from a to b, then

$$\mathfrak{p}^t \colon a + t = a_0 + t \xrightarrow{m_1} a_1 + t \longrightarrow \cdots \xrightarrow{m_k} a_k + t = b + t$$

is a path from $a + t$ to $b + t$.

Every path is now a sum of edges, and every edge $a \xrightarrow{m} b$ is a translate of a simple path $m \longrightarrow r_m$. Thus, addition and translation build all paths from simple paths.

6.1.4 Overpaths

Relative to a compatible well order on $\mathbb{F}$, an *overpath* in $\mathbb{F}$ is a sequence of elements of M.

If $\mathfrak{p}: a = a_0 \xrightarrow{m_1} a_1 \longrightarrow \cdots \xrightarrow{m_n} a_n = b$ is a path from a to b, then the sequence $P = (m_1, \ldots, m_n)$ is the *overpath* of $\mathfrak{p}$, and is an overpath *from a to b*. Like the stars in the sky the overpath of $\mathfrak{p}$ guides weary travelers in their trek through $\mathbb{F}$ from a to b:

$$
\begin{array}{ccccccccc}
P: & & m_1 & & m_2 & & \cdots & & m_n \\
\mathfrak{p}: & a = a_0 & \longrightarrow & a_1 & \longrightarrow & a_2 & \longrightarrow & \cdots & \longrightarrow & a_n = b
\end{array}
$$

A path and its translates all share the same overpath.

A path is completely determined by its origin and its overpath: if

$$
\mathfrak{p}: a = a_0 \xrightarrow{m_1} a_1 \longrightarrow \cdots \xrightarrow{m_n} a_n = b
$$

is a path from a to b, then $a_{i-1} - a_i = \overrightarrow{m_i}$ for all $1 \leq i \leq n$, so that

$$
a_j = a - \sum_{1 \leq i \leq j} \overrightarrow{m_i} \quad \text{and} \quad b = a - \sum_{1 \leq i \leq n} \overrightarrow{m_i}, \tag{6.1}
$$

as calculated in $\mathbb{G}$. Conversely, if $a \in \mathbb{F}$, then $P = (m_1, \ldots, m_n)$ is the overpath of a path from a if and only if

$$
a_j = a - \sum_{1 \leq i \leq j} \overrightarrow{m_i} \in \mathbb{F} \quad \text{and} \quad a_j \geq m_j \tag{6.2}
$$

for all $1 \leq j \leq n$; then

$$
a = a_0 \xrightarrow{m_1} a_1 \longrightarrow \cdots \xrightarrow{m_n} a_n = b
$$

is a path from a to $b = a - \sum_{1 \leq i \leq n} \overrightarrow{m_i}$ with overpath P.

6.2 Main Result

We now apply the concepts in Sect. 6.1 to the cohomology of $S \cong \mathbb{F}/\mathcal{C}$.

Let $\pi: \mathbb{F} \longrightarrow S$ be the projection. Every abelian group valued functor $\mathcal{G} = (G, \gamma)$ on S lifts to an abelian group valued functor $\mathcal{G}^* = (G^*, \gamma^*)$ on $\mathbb{F}$ that assigns $G_a^* = G_{\pi a}$ to every $a \in \mathbb{F}$ and $\gamma_{a,t}^* = \gamma_{\pi a, \pi t}: G_a^* \longrightarrow G_{a+t}^*$ to every $a \in \mathbb{F}$

and $t \in \mathbb{F}$ (so that $g^t = g^{\pi t}$). Note that $\mathcal{G}^*$ is always thin (even if $\mathcal{G}$ is not): if $a + t = a + u$, then cancellation in $\mathbb{F}$ yields $t = u$ and $\gamma_{a,t}^* = \gamma_{a,u}^*$.

In what follows we denote $G_{\pi a}$ by G_a and $\gamma_{\pi a, \pi t}$ by $\gamma_{a,t}$, or by γ_{a+t}^a. This brazen abuse of language should not create confusion, since it will be clear from context whether a and t are in S or in $\mathbb{F}$. Thus γ_b^a is defined whenever $\pi a \geq \pi b$ in S, in particular whenever $a \leq b$ in $\mathbb{F}$.

6.2.1 Minimal Cocycles

A *minimal* [1-]*cochain* on S with values in $\mathcal{G}$ is a family $u = (u(m))$ $(m \in M)$ such that $u(m) \in G_m$ for all $m \in M$. Under pointwise addition, minimal cochains constitute an abelian group $C_m(S, \mathcal{G})$.

Given $u \in C_m(S, \mathcal{G})$ and a path

$$\mathfrak{p}: \ a = a_0 \xrightarrow{m_1} a_1 \xrightarrow{m_2} a_2 \longrightarrow \cdots \longrightarrow a_{n-1} \xrightarrow{m_n} a_n = b \,,$$

let

$$u(\mathfrak{p}) = \sum_{1 \leq i \leq n} \gamma_a^{m_i} u(m_i) \in G_a \,,$$

where $\gamma_a^{m_i}$ is defined since $a_i \geq m_i$ and $\pi a_i = \pi a$. Note that $u(\mathfrak{p})$ depends only on the overpath of $\mathfrak{p}$.

Also note the following properties. If $\mathfrak{p}: m \longrightarrow r_m$ is a simple path, then $u(\mathfrak{p}) = u(m)$. If $\mathfrak{p}: a \longrightarrow b$ and $\mathfrak{q}: b \longrightarrow c$, then $\pi a = \pi b = \pi c$ and

$$u(\mathfrak{p} + \mathfrak{q}) = u(\mathfrak{p}) + u(\mathfrak{q}) \,.$$

If $t \in \mathbb{F}$, then

$$u(\mathfrak{p}^t) = \gamma_{a+t}^a \, u(\mathfrak{p}) = u(\mathfrak{p})^t \,.$$

If $u, v \in C_m(S, \mathcal{G})$, then

$$(u + v)(\mathfrak{p}) = u(\mathfrak{p}) + v(\mathfrak{p}) \,.$$

A *minimal cocycle* on S with values in $\mathcal{G}$ is a minimal cochain u such that $u(\mathfrak{p}) = u(\mathfrak{q})$ whenever $\mathfrak{p}, \mathfrak{q}: a \longrightarrow r_a$ are paths from a to r_a. Under pointwise addition, minimal cocycles constitute a subgroup $Z_m(S, \mathcal{G})$ of $C_m(S, \mathcal{G})$.

If u is a minimal cocycle, then $u(\mathfrak{p}) = u(\mathfrak{q})$ whenever $\mathfrak{p}, \mathfrak{q}: a \longrightarrow b$: since there exists a path $\mathfrak{r}: b \longrightarrow r_b = r_a$, we have

$$u(\mathfrak{p}) + u(\mathfrak{r}) = u(\mathfrak{p} + \mathfrak{r}) = u(\mathfrak{q} + \mathfrak{r}) = u(\mathfrak{q}) + u(\mathfrak{r})$$

and $u(\mathfrak{p}) = u(\mathfrak{q})$.

If u is a minimal cocycle, then $u(a)$ is well-defined for every $a \in \mathbb{F}$ by: $u(a) = u(\mathfrak{p})$ whenever $\mathfrak{p}$ is an overpath from a to r_a. This extends every minimal cocycle to a 1-cochain on $\mathbb{F}$.

6.2.2 Main Result

Let $w = (w(x))$ $(x \in X)$ be a 1-cochain on X, so that $w(x) \in G_x$ for all $x \in X$. Define a minimal cochain δw by

$$(\delta w)(m) = \sum_{x \in X,\, \pi x \geq \pi m} \overrightarrow{m}_x\, \gamma_m^x\, w(x) \in G_m \tag{6.3}$$

for every $m = \sum_{x \in X} m_x\, x \in M$ (where $\overrightarrow{m} = m - r_m$).

A *minimal coboundary* is a minimal cochain δw calculated by (6.3) for some 1-cochain $w = (w(x))$ $(x \in X)$ on X.

Under pointwise addition, minimal coboundaries constitute a subgroup $B_m(S, \mathcal{G})$ of $C_m(S, \mathcal{G})$.

Lemma 6.2.1 *Every minimal coboundary is a minimal cocycle.*

Proof The proof in [22] and [31] is not quite satisfactory, so we give the proof from [33]. Let $\mathfrak{p}$ be a path from a to r_a with overpath $(m_1, \ldots, m_n)$. Then $a - r_a = \sum_{1 \leq i \leq n} \overrightarrow{m}_i$,

$$a_x - (r_a)_x = \sum_{1 \leq i \leq n} (\overrightarrow{m}_i)_x \tag{6.4}$$

for all $x \in X$, and

$$(\delta w)(\mathfrak{p}) = \sum_{1 \leq i \leq n} \gamma_a^{m_i}\, (\delta w)(m_i)$$

$$= \sum_{1 \leq i \leq n} \gamma_a^{m_i} \Big(\sum_{x \in X,\, \pi x \geq \pi m_i} (\overrightarrow{m}_i)_x\, \gamma_{m_i}^x\, w(x) \Big)$$

$$= \sum_{1 \leq i \leq n} \sum_{x \in X,\, \pi x \geq \pi m_i} (\overrightarrow{m}_i)_x\, \gamma_a^x\, w(x).$$

Now $\pi m_i \geq \pi a$ for all i, so that $\pi x \geq \pi m_i$ implies $\pi x \geq \pi a$. If $\pi x \geq \pi a$ but $\pi x \not\geq \pi m_i = \pi r_{m_i}$, then $x \not\leq m_i$, $x \not\leq r_{m_i}$, $(m_i)_x = (r_{m_i})_x = 0$, and $(\overrightarrow{m_i})_x = 0$. Hence

$$
\begin{aligned}
(\delta w)(\mathfrak{p}) &= \sum_{1 \leq i \leq n} \sum_{x \in X, \pi x \geq \pi a} (\overrightarrow{m_i})_x \, \gamma_a^x \, w\,(x) \\
&= \sum_{x \in X, \pi x \geq \pi a} \sum_{1 \leq i \leq n} (\overrightarrow{m_i})_x \, \gamma_a^x \, w\,(x) \\
&= \sum_{x \in X, \pi x \geq \pi a} (a_x - (r_a)_x) \, \gamma_a^x \, w\,(x) \,,
\end{aligned}
$$

by (6.4). Thus $(\delta w)(\mathfrak{p})$ depends only on a, and not on the choice of $\mathfrak{p}$.   □

By Lemma 6.2.1, $B_m(S, \mathcal{G})$ is a subgroup of $Z_m(S, \mathcal{G})$. The main result of [22] and [26] is:

Theorem 6.2.1 *There is an isomorphism*

$$
H^2(S, \mathcal{G}) \cong Z_m(S, \mathcal{G}) \,/\, B_m(S, \mathcal{G})
$$

which is natural in $\mathcal{G}$.

Theorem 6.2.1 is proved as follows [31]. A symmetric 2-cocycle $s \in Z_S^2(S, \mathcal{G})$ on S lifts through the projection $\mathbb{F} \longrightarrow S$ to a symmetric 2-cocycle $s^* \in Z_S^2(\mathbb{F}, \mathcal{G}^*)$ on $\mathbb{F}$. Since $H^2(\mathbb{F}, \mathcal{G}^*) = 0$, s^* is the coboundary of a 1-cochain $u \in C^1(\mathbb{F}, \mathcal{G}^*)$. Then u is repeatedly trimmed down to a minimal cocycle.

6.2.3 Examples

We begin with the groupfree semigroup of Example 1. As Example 2 in Chap. 3 we determined that $H^2(S, \mathcal{G}) = 0$ when G_1 and G_a are cyclic of order 2 and $G_e = G_0 = 0$. To illustrate the overpath method we now compute $H^2(S, \mathcal{G})$ for *any* thin abelian group valued functor $\mathcal{G}$ on S.

Example 3 Example 2 is $S = \{1, a, e, 0\}$, with multiplication table

$$
\begin{array}{|llll}
1 & a & e & 0 \\
a & 0 & 0 & 0 \\
e & 0 & e & 0 \\
0 & 0 & 0 & 0
\end{array}
$$

As a commutative monoid with a zero element, S is generated by a and e subject to $a^2 = ae = 0$ and $e^2 = e$. As a commutative monoid, S is generated by a and e

subject to $e^2 = e$ and $a^2 = ae = aea = ae^2$, these last equalities ensuring that ae is a zero element.

Hence $S \cong \mathbb{F}/\mathcal{C}$, where $\mathbb{F}$ has two generators $x = (1, 0)$ and $y = (0, 1)$, with $\pi x = e$ and $\pi y = a$, and the four $\mathcal{C}$-class es I, A, E, Z are shown below:

$$
\begin{array}{c|cccc}
3 & m & & & \\
 & & & Z & \\
2 & r_Z & & & \\
1 & A & n & & \\
0 & I & r_E & p & E \\
\hline
 & 0 & 1 & 2 & 3
\end{array}
$$

Put the lexicographic order on $\mathbb{F}$ induced by $x < y$, so that $(q, r) \prec (s, t)$ in $\mathbb{F}$ if either $q < s$, or $q = s$ and $r < t$. The set R consists of

$$r_I = (0, 0), \quad r_A = (0, 1), \quad r_E = (1, 0), \quad \text{and} \quad r_Z = (0, 2) \,.$$

Hence M has three elements:

$$m = (0, 3), \quad n = (1, 1), \quad \text{and} \quad p = (2, 0),$$

where $m, n \in Z$ and $p \in E$, with defining vectors

$$\overrightarrow{m} = (0, 1), \quad \overrightarrow{n} = (1, -1), \quad \text{and} \quad \overrightarrow{p} = (1, 0).$$

If u is a minimal cochain, then $u(m), u(n) \in G_0$ and $u(p) \in G_e$; hence $C_m(S, \mathcal{G}) = G_0 \oplus G_0 \oplus G_e$.

Next we find the minimal cocycles. For any minimal cochain u independence of path is ensured in I and A, which have only empty paths, and in E, since there is only one path from $a \in E$ to r_E. The $\mathcal{C}$-class Z is more complicated.

Hindsight from the next section yields two paths in Z from $(2, 1)$ to $(1, 1)$:

$$\mathfrak{p} \colon (2, 1) \xrightarrow{\ n\ } (1, 2) \xrightarrow{\ m\ } (1, 1) \quad \text{and} \quad \mathfrak{q} \colon (2, 1) \xrightarrow{\ p\ } (1, 1).$$

If u is a minimal cocycle, then $u(\mathfrak{p}) = u(\mathfrak{q})$ and

$$u(m) + u(n) = \gamma_0^e \, u(p). \tag{6.5}$$

Conversely, assume that (6.5) holds and let $\mathfrak{p}$ be any path from $a \in Z$ to $r_Z = (0, 2)$. Let the overpath of $\mathfrak{p}$ consists of i copies of m, j copies of n, and k copies of p (in any order), so that

$$a - r_Z = i \overrightarrow{m} + j \overrightarrow{n} + k \overrightarrow{p} = (j + k, \ i - j)$$

and

$$u(\mathfrak{p}) \;=\; i\,u(m) + j\,u(n) + k\,\gamma_0^e\,u(p) \;=\; (i+k)\,u(m) + (j+k)\,u(n).$$

Now $a - r_Z = (j+k,\ i-j)$ does not depend on the choice of $\mathfrak{p}$, hence $j+k$ does not depend on the choice of $\mathfrak{p}$, and neither does $i+k = (j+k)+(i-j)$. Therefore u is a minimal cocycle.

If u is a minimal cocycle, then (6.5) determines $u(n)$ from $u(m)$ and $u(p)$. Therefore the projection $\rho: u \longmapsto (u(m), u(p))$ of $C_m(S, \mathcal{G})$ onto $G_0 \oplus G_e$ induces an isomorphism $Z_m(S, \mathcal{G}) \cong G_0 \oplus G_e$.

Finally we find the minimal coboundaries. Let w be a 1-cochain on $X = \{x, y\}$, so that $g = w(x) \in G_e$ and $h = w(y) \in G_a$. Since $\pi m = 0$ we have $\pi x, \pi y \geq \pi m$, and similarly for n. Hence (6.3): $(\delta w)(m) = \sum_{x \in X,\, \pi x \geq \pi m} \overrightarrow{m_x}\, \gamma_m^x\, w(x)$ yields

$$(\delta w)(m) \;=\; \gamma_m^y\, w(y) \;=\; \gamma_0^a\, h,$$

$$(\delta w)(n) \;=\; \gamma_n^x\, w(x) \;-\; \gamma_n^y\, w(y) \;=\; \gamma_0^e\, g \;-\; \gamma_0^a\, h,$$

$$(\delta w)(p) \;=\; \gamma_p^x\, w(x) \;=\; g.$$

Therefore the map $\rho: u \longmapsto (u(m), u(p))$ above induces an isomorphism $B_m(S, \mathcal{G}) \cong \operatorname{Im} \gamma_0^a \oplus G_e$.

Hence $H^2(S, \mathcal{G}) \cong G_0 / \operatorname{Im} \gamma_0^a$.

In particular, $H^2(S, \mathcal{G}) = 0$ if $G_0 = 0$, as in Example 2 in Chap. 3.

6.2.4 Semigroups with One Relator

Our next example is a commutative monoid with one defining relation

$$S \;=\; \langle\, a_1, \ldots, a_n, \cdots \mid a_1^{p_1} a_2^{p_2} \cdots a_n^{p_n} \;=\; a_1^{q_1} a_2^{q_2} \cdots a_n^{q_n} \,\rangle,$$

where all $p_i, q_i \geq 0$ are integers, $p_i \neq q_i$ for some i, and we may assume that $p_i + q_i > 0$ for all i. Let $\mathcal{G}$ be any abelian group valued functor on S.

Let $\mathbb{F}$ be the free commutative monoid on $X = \{x_1, \ldots, x_n, \ldots\}$, where $x_1, \ldots, x_n, \ldots$ are distinct, so that $\pi: \mathbb{F} \longrightarrow S$ sends x_i to a_i, and let $\prec$ be any compatible well order on $\mathbb{F}$. Since

$$p \;=\; \sum_{1 \leq i \leq n} p_i x_i \quad \text{and} \quad q \;=\; \sum_{1 \leq i \leq n} q_i x_i$$

are distinct, we may assume that $p \succ q$. Also, $\mathcal{C}$ is generated by (p, q), so that p is the sole element of M, with $r_p = q$,

$$\vec{p} = p - q = \sum_{1 \le i \le n} (p_i - q_i) x_i ,$$

and there is only one path from any $a \in \mathbb{F}$ to r_a, whose overpath is a sequence of p's. Therefore $Z_m(S, \mathcal{G}) = C_m(S, \mathcal{G}) = G_d$, where

$$d = a_1^{p_1} a_2^{p_2} \cdots a_n^{p_n} = a_1^{q_1} a_2^{q_2} \cdots a_n^{q_n} = \pi p = \pi q.$$

Let w be a 1-cochain on X and let $w(x_i) = w_i \in G_{a_i}$. By (6.3),

$$(\delta w)(p) = \sum_{x \in X, \, \pi x \ge \pi p} (p - q)_x \, \gamma_p^x \, w(x)$$

$$= \sum_{i \le n, \, p_i > 0} p_i \, \gamma_p^{x_i} w_i - \sum_{i \le n, \, q_i > 0} q_i \, \gamma_q^{x_i} w_i . \tag{6.6}$$

Let

$$d_i' = a_1^{p_1} a_2^{p_2} \cdots a_{i-1}^{p_{i-1}} a_i^{p_i-1} a_{i+1}^{p_{i+1}} \cdots a_n^{p_n}$$

if $p_i > 0$, so that $a_i \, d_i' = d$ and $\gamma_p^{x_i} = \gamma_{a_i, d_i'}$, and let

$$d_i'' = a_1^{q_1} a_2^{q_2} \cdots a_{i-1}^{q_{i-1}} a_i^{q_i-1} a_{i+1}^{q_{i+1}} \cdots a_n^{q_n}$$

if $q_i > 0$, so that $a_i \, d_i'' = d$ and $\gamma_q^{x_i} = \gamma_{a_i, d_i''}$, and let $\gamma_i : G_{a_i} \longrightarrow G_d$ be the following homomorphism:

$$\gamma_i = \begin{cases} p_i \, \gamma_{a_i, d_i'} - q_i \, \gamma_{a_i, d_i''} & \text{if } p_i, q_i > 0, \\ p_i \, \gamma_{a_i, d_i'} & \text{if } p_i > 0 \text{ and } q_i = 0, \\ -q_i \, \gamma_{a_i, d_i''} & \text{if } p_i = 0 \text{ and } q_i > 0. \end{cases}$$

Then (6.6) yields $B_m(S, \mathcal{G}) = \sum_{1 \le i \le n} \text{Im} \, \gamma_i$ and we obtain

Proposition 6.2.1 *If*

$$S = \langle a_1, \ldots, a_n, \cdots \mid a_1^{p_1} a_2^{p_2} \cdots a_n^{p_n} = a_1^{q_1} a_2^{q_2} \cdots a_n^{q_n} \rangle$$

is a one-relator commutative monoid, then

$$H^2(S, \mathcal{G}) \cong G_d \Big/ \sum_{1 \le i \le n} \operatorname{Im} \gamma_i \,,$$

where d and γ_i are as above.

In particular,

Corollary 1 *If $S = \langle a \mid a^r = a^{r+p} \rangle$ is cyclic of index r and period p, then $H^2(S, \mathcal{G}) \cong G_{a^r} / p \operatorname{Im} \gamma_{a, a^{r-1}}$.*

6.3 Other Results

This section contains additional enhancements and applications of the Overpath Method.

6.3.1 Branching

Branching occurs when there is more than one path from a to r_a.

The following definitions were introduced in [33] for nilsemigroups but are presented here in full generality, with some simplifications. Recall that $\mathbb{F}$ is a lattice under $\le$, in which $a \vee b = c$ has $c_x = \max(a_x, b_x)$ for all $x \in X$. A *branch* occurs at $a \in \mathbb{F}$, and a is a *branching point*, if and only if there exist $m, n \in M$ such that $m \ne n$ and $a = m \vee n$. Then there exists a pair $(\mathfrak{p}, \mathfrak{q})$ of paths from a to r_a, in which the overpath of $\mathfrak{p}$ begins with m and the overpath of $\mathfrak{q}$ begins with n:

$$
\begin{array}{ccc}
m & \overset{m}{\longrightarrow} & r_m \\
\vdots & & \vdots \\
\mathfrak{p} = \quad a & \overset{m}{\longrightarrow} & a - \overrightarrow{m} \longrightarrow \cdots \longrightarrow r_a, \quad
\end{array}
\qquad
\begin{array}{ccc}
n & \overset{n}{\longrightarrow} & r_n \\
\vdots & & \vdots \\
\mathfrak{q} = \quad a & \overset{n}{\longrightarrow} & a - \overrightarrow{n} \longrightarrow \cdots \longrightarrow r_a.
\end{array}
$$

For each $m, n \in M$ such that $m \ne n$ and $a = m \vee n$, select one such pair $(\mathfrak{p}, \mathfrak{q})$ to be the corresponding *branching pair at a*. (It is not necessary to choose a path from a to r_a for *every* $a \in \mathbb{F}$, as in [33].) A minimal cocycle must then satisfy the *branching condition*: $u(\mathfrak{p}) = u(\mathfrak{q})$, for every branching pair $(\mathfrak{p}, \mathfrak{q})$, which, like $u(\mathfrak{p})$, depends only on the overpaths of $\mathfrak{p}$ and $\mathfrak{q}$.

Branching pairs generate all branching. Hence the next result, which is essentially Lemma 2.5 of [33].

Lemma 6.3.1 *A minimal cochain is a minimal cocycle if and only if it satisfies all branching conditions.*

Proof Assume that $u(\mathfrak{p}) = u(\mathfrak{q})$ for all branching pairs $(\mathfrak{p}, \mathfrak{q})$ and let $\mathfrak{p}, \mathfrak{q}$ be any paths from $a \in \mathbb{F}$ to r_a. That $u(\mathfrak{p}) = u(\mathfrak{q})$ is proved by well-ordered induction on a.

First, $u(\mathfrak{p}) = u(\mathfrak{q})$ holds vacuously if $a = r_a$. Let $a \neq r_a$ and let the overpaths of $\mathfrak{p}$ and $\mathfrak{q}$ begin with m and n. Let $b = a - \overrightarrow{m}$ and $c = a - \overrightarrow{n}$, so that $m, n \leq a$, $\mathfrak{p} = (a \xrightarrow{m} b) + \mathfrak{p}'$, and $\mathfrak{q} = (a \xrightarrow{n} c) + \mathfrak{q}'$, where $\mathfrak{p}': b \longrightarrow r_b = r_a$ and $\mathfrak{q}': c \longrightarrow r_c = r_a$.

(a) If $m = n$, then $b = c \prec a$ and the induction hypothesis yields $u(\mathfrak{p}') = u(\mathfrak{q}')$ and $u(\mathfrak{p}) = u(m) + u(\mathfrak{p}') = u(n) + u(\mathfrak{q}') = u(\mathfrak{q})$.

(b) If $m \neq n$ and $a = m \vee n$, then there is a branching pair $(\mathfrak{r}, \mathfrak{s})$ of paths from a to r_a, in which $\mathfrak{r} = (a \xrightarrow{m} b) + \mathfrak{r}'$ and $\mathfrak{s} = (a \xrightarrow{n} c) + \mathfrak{s}'$. Then $u(\mathfrak{r}) = u(\mathfrak{s})$ and (a) yields $u(\mathfrak{p}) = u(\mathfrak{r}) = u(\mathfrak{s}) = u(\mathfrak{q})$.

(c) If $m \neq n$ and $a > d = m \vee n$, then there is a branching pair $(\mathfrak{r}, \mathfrak{s})$ of paths from d to r_d, where $\mathfrak{r} = (d \xrightarrow{m} e) + \mathfrak{r}'$ and $\mathfrak{s} = (d \xrightarrow{n} f) + \mathfrak{s}'$ for some $e, f \in \mathbb{F}$. Then $u(\mathfrak{r}) = u(\mathfrak{s})$.

Now $a = d + t$ for some $t \in \mathbb{F}$; translating $\mathfrak{r}$, $\mathfrak{s}$ yields $\mathfrak{r}^t$, $\mathfrak{s}^t: a = d + t \longrightarrow r_d + t$, and

$$u(\mathfrak{r}^t) = \gamma_a^d \, u(\mathfrak{r}) = \gamma_a^d \, u(\mathfrak{s}) = u(\mathfrak{s}^t).$$

Since $r_d + t \in C_a$, there exists a path $\mathfrak{t}: r_d + t \longrightarrow r_a$. Then the overpath of $\mathfrak{r}^t + \mathfrak{t}: a \longrightarrow r_a$ begins with m, the overpath of $\mathfrak{s}^t + \mathfrak{t}: a \longrightarrow r_a$ begins with n, and (a) yields $u(\mathfrak{p}) = u(\mathfrak{r}^t + \mathfrak{t})$ and $u(\mathfrak{q}) = u(\mathfrak{s}^t + \mathfrak{t})$. Hence

$$u(\mathfrak{p}) = u(\mathfrak{r}^t + \mathfrak{t}) = u(\mathfrak{r}^t) + u(\mathfrak{t}) = u(\mathfrak{s}^t) + u(\mathfrak{t}) = u(\mathfrak{s}^t + \mathfrak{t}) = u(\mathfrak{q}).$$

Thus u is a minimal cocycle. $\square$

Branching and Lemma 6.3.1 limit the number of paths that need to be considered when determining minimal cocycles. We illustrate this with another look at Example 3.

Example 4 In Example 3 we saw that M has three elements:

$$m = (0, 3) \in Z, \quad n = (1, 1) \in Z, \quad \text{and} \quad p = (2, 0) \in E,$$

with defining vectors

$$\overrightarrow{m} = (0, 1), \quad \overrightarrow{n} = (1, -1), \quad \text{and} \quad \overrightarrow{p} = (1, 0).$$

There are three branching points:

$$m \vee n = (1, 3), \quad m \vee p = (2, 3), \quad \text{and} \quad n \vee p = (2, 1).$$

We have $(1, 3) - \vec{m} = (1, 2)$ and $(1, 3) - \vec{n} = (0, 4)$. Select

$$\mathfrak{p}\colon (1, 3) \xrightarrow{\ m\ } (1, 2) \xrightarrow{\ m\ } (1, 1) \xrightarrow{\ n\ } (0, 2) \quad \text{and}$$

$$\mathfrak{q}\colon (1, 3) \xrightarrow{\ n\ } (0, 4) \xrightarrow{\ m\ } (0, 3) \xrightarrow{\ m\ } (0, 2),$$

as the corresponding branching pair. The resulting branching condition is trivial.
  Similarly, $(2, 3) - \vec{m} = (2, 2)$ and $(2, 3) - \vec{p} = (1, 3)$; select

$$\mathfrak{p}\colon (2, 3) \xrightarrow{\ m\ } (2, 2) \xrightarrow{\ p\ } (1, 2) \xrightarrow{\ p\ } (0, 2) \quad \text{and}$$

$$\mathfrak{q}\colon (2, 3) \xrightarrow{\ p\ } (1, 3) \xrightarrow{\ m\ } (1, 2) \xrightarrow{\ m\ } (1, 1) \xrightarrow{\ n\ } (0, 2),$$

as the corresponding branching pair. The resulting branching condition is: $u(m) + 2\gamma_0^e u(p) = \gamma_0^e u(p) + 2u(m) + u(n)$, equivalently

$$\gamma_0^e u(p) = u(m) + u(n), \tag{6.7}$$

which reflects the equality $\vec{p} = \vec{m} + \vec{n}$ and we recognize as (6.5).
  Finally, $(2, 1) - \vec{n} = (1, 2)$ and $(2, 1) - \vec{p} = (1, 1)$; select

$$\mathfrak{p}\colon (2, 1) \xrightarrow{\ n\ } (1, 2) \xrightarrow{\ m\ } (1, 1) \xrightarrow{\ n\ } (0, 2) \quad \text{and}$$

$$\mathfrak{q}\colon (2, 1) \xrightarrow{\ p\ } (1, 1) \xrightarrow{\ n\ } (0, 2),$$

as the corresponding branching pair. The resulting branching condition is the same as (6.7):

$$\gamma_0^e u(p) = u(m) + u(n).$$

By Lemma 6.3.1, a minimal cochain u is a minimal cocycle if and only if it has property (6.7).

6.3.2   Relations

The *relations* in question are linear relations between defining vectors, with integer coefficients:

$$\sum_{m \in M} r_m \vec{m} = 0.$$

Every relation can be rewritten as a *positive relation*: an equality between two linear combinations with nonnegative integer coefficient s:

$$R(p, q): \sum_{m \in M} p_m \overrightarrow{m} = \sum_{m \in M} q_m \overrightarrow{m}.$$

Relations enable branching: if $\mathfrak{p}$ and $\mathfrak{q}$ are paths in $\mathbb{F}$ from a to b, with overpaths $(m_1, \ldots, m_k)$ and $(n_1, \ldots, n_\ell)$, then

$$R(\mathfrak{p}, \mathfrak{q}): \sum_{1 \le i \le k} \overrightarrow{m}_i = a - b = \sum_{1 \le j \le \ell} \overrightarrow{n}_j.$$

is a relation between defining vectors, without which $\mathfrak{p}$ and $\mathfrak{q}$ would have different destinations. If $p(m)$ (resp. $q(m)$) is the number of appearances of m in the overpath of $\mathfrak{p}$ (resp. $\mathfrak{q}$), then $R(\mathfrak{p}, \mathfrak{q}) = R(p, q)$. Thus, minimal cocycles are determined by relations $R(\mathfrak{p}, \mathfrak{q})$.

A relation is *realized* at $a \in \mathbb{F}$ (resp. in a $\mathcal{C}$-class C) if and only if it is a trivial consequence of $R(\mathfrak{p}, \mathfrak{q})$ for some paths $\mathfrak{p}$ and $\mathfrak{q}$ from a to the same $b \in \mathbb{F}$ (resp. from some $a \in C$ to the same $b \in C$). (A trivial consequence is obtained by adding terms to both sides, and/or canceling terms from both sides.)

In Example 3 we used the fact that the relation $\overrightarrow{m} + \overrightarrow{n} = \overrightarrow{p}$ is realized in the $\mathcal{C}$-class Z.

Proposition 6.3.1 *Every relation between defining vectors is realized in some $\mathcal{C}$-class (in the zero class, if S has a zero element).*

Proof This is Prop. XIII.4.5 of [31]. Given a positive relation $\sum_{m \in M} p_m \overrightarrow{m} = \sum_{m \in M} q_m \overrightarrow{m}$ between defining vectors, construct a sequence $(m_1, \ldots, m_k)$ that consists of p_m copies of each $m \in M$; this works since $\{m \in M \mid p_m > 0\}$ is finite in the linear combination $\sum_{m \in M} p_m \overrightarrow{m}$. Similarly there is a sequence $(n_1, \ldots, n_\ell)$ that consists of q_m copies of each $m \in M$.

If $a \in \mathbb{F}$ has sufficiently high coordinates, then

$$a_j = a - \sum_{1 \le i \le j} \overrightarrow{m}_i \in \mathbb{F}$$

for all $j \le k$. In particular $c = a_k \in \mathbb{F}$. Similarly

$$b_j = a - \sum_{1 \le i \le j} \overrightarrow{n}_i \in \mathbb{F}$$

for all $j \le \ell$. In particular $b_\ell \in \mathbb{F}$. The given relation ensures that

$$b_\ell = a - \sum_{m \in M} q_m \overrightarrow{m} = a - \sum_{m \in M} p_m \overrightarrow{m} = a_k.$$

Hence $(m_1, \ldots, m_k)$ and $(n_1, \ldots, n_\ell)$ are the overpaths of two paths $\mathfrak{p}$ and $\mathfrak{q}$ from a to c, and the given relation is realized in the $\mathcal{C}$-class of a.

If S has a zero element and a has sufficiently high coordinates, then a, c, and every a_j and b_j all belong to the zero class Z, and the given relation is realized in Z.   □

Minimal cocycles are determined by relations that are realized:

Proposition 6.3.2 *A minimal cochain u is a minimal cocycle if and only if*

$$\sum_{m \in M, \pi m \geq \pi a} r_m \, \gamma_a^m u(m) = 0 \tag{6.8}$$

whenever $a \in \mathbb{F}$ and $\sum_{m \in M} r_m \overrightarrow{m} = 0$ is realized at a.

In Example 3 the condition (6.7): $\gamma_0^e u(p) = u(m) + u(n)$ arises in this fashion from the relation $\overrightarrow{p} = \overrightarrow{m} + \overrightarrow{n}$, which is realized at any $a \in Z$.

A *defining vector basis* is a basis of $\mathbb{G}$ (the free abelian group on X) that consists of defining vectors. For instance this is the case in Example 3, where $\overrightarrow{m} = (0, 1)$ and $\overrightarrow{p} = (1, 0)$ constitute a defining vector basis of $\mathbb{G}$. This facilitates the determination of $H^2(S, \mathcal{G})$.

A defining vector basis of $\mathbb{G}$ is determined by the elements of M whose defining vectors constitute it.

Proposition 6.3.3 *There exists a defining vector basis if and only if there exists a subset L of M with the following properties:*

(a) $\{\overrightarrow{m} \mid m \in L\}$ *is a basis of $\mathbb{G}$; and*
(b) *if $m, n \in L$ and $m \neq n$, then $\overrightarrow{m} \neq \overrightarrow{n}$.*

Assume that there is a defining vector basis and let L be as in Proposition 6.3.3. For each $m \in M \backslash L$ there exist unique integers $k_{m,\ell}$ such that

$$\overrightarrow{m} = \sum_{\ell \in L} k_{m,\ell} \overrightarrow{\ell}. \tag{6.9}$$

We refer to (6.9) as a *basic relation*. In Example 3 the relation $\overrightarrow{n} = \overrightarrow{m} - \overrightarrow{p}$ is a basic relation. In general, if there is a defining vector basis, then all existing relations between defining vectors are consequences of the basic relations.

If the basic relation (6.9) is realized at $a \in \mathbb{F}$, then Proposition 6.3.2 yields a *basic condition*

$$\gamma_a^m u(m) = \sum_{\ell \in L, \pi \ell \geq \pi a} k_{m,\ell} \, \gamma_a^\ell u(\ell). \tag{6.10}$$

In Example 3 the condition (6.7): $u(n) = u(m) - \gamma_0^e u(p)$ is a basic condition.

Proposition 6.3.4 *If S has a zero element and a defining vector basis, then a minimal cochain u satisfies $u(\mathfrak{p}) = u(\mathfrak{q})$ for all coterminal paths $\mathfrak{p}$ and $\mathfrak{q}$ within the zero class if and only if it satisfies all basic conditions.*

This is essentially Lemma 2.1 of [30]. For example, in Example 3, the condition $u(n) = u(m) - \gamma_0^e u(p)$ ensures $u(\mathfrak{p}) = u(\mathfrak{q})$ for all coterminal paths $\mathfrak{p}$ and $\mathfrak{q}$ within Z.

6.3.3 Partially Free Semigroups

A *nilmonoid* is a commutative nilsemigroup to which has been adjoined an identity element. A nilmonoid is *0-free* if and only if it is the Rees quotient F/I of a free commutative monoid F by an ideal I.

An early result of Prostakovsky [43] implies that every finite groupfree commutative monoid has a canonical subdirect decomposition into finitely many nilmonoids. A *partially free* monoid is a finite groupfree commutative monoid such that all nilmonoids in its canonical subdirect decomposition are 0-free. This definition, and the more general but more technical definition in [27] and [31], arose from the detailed analysis of congruences in [23].

For example a finite nilmonoid is partially free if and only if it is 0-free. Example 3, $S = \{1, a, e, 0\}$ with multiplication table

	1	a	e	0
1	a	e	0
a	0	0	0
e	0	e	0
0	0	0	0

is partially free: the projection $S \longrightarrow N_1 = S/\{e, 0\} \cong \{1, a, 0\}$ and multiplication by $e: S \longrightarrow N_2 = \{e, 0\}$ yield the canonical subdirect decomposition of S, in which N_1 and N_2 are singly generated nilmonoids, hence 0-free.

The following result shows that partially free commutative monoids constitute a fairly large class. A monoid homomorphism $\varphi: S \longrightarrow T$ is *pure* if and only if the inverse image $\varphi^{-1}(e)$ of every idempotent e of T has at most one element (necessarily an idempotent).

Proposition 6.3.5 *For every finite groupfree commutative monoid S there exists a pure homomorphism of a partially free monoid onto S.*

This is Prop. X.6.2 of [31] and was first noted for nilmonoids in [4].

As of this writing, partially free commutative monoids constitute the only relatively large class of commutative semigroups with known cohomology. Let S be a partially free commutative monoid. An element c of S is *irreducible* if and only if $c \neq 1$ and $c = xy$ implies $x = c$ or $y = c$.

Theorem 6.3.1 *If S is a finite partially free commutative monoid and $\mathcal{G}$ is thin, then there is an isomorphism which is natural in $\mathcal{G}$*

$$H^2(S, \mathcal{G}) \cong \bigoplus_{c \in Irr(S)} G_{e(c)}/\operatorname{Im} \gamma_{e(c)}^c,$$

where $Irr(S)$ is the set of all irreducible elements of S and $e(c)$ denote the idempotent e of S which is also a power of c.

For instance, Example 3 is partially free, with one irreducible element b, and we found $H^2(S, \mathcal{G}) \cong G_0 / \operatorname{Im} \gamma_0^b$ if $\mathcal{G}$ is thin.

Theorem 6.3.1 is Theorem XIII.5.6 of Bk, from [27]. The proof uses technical properties of $\mathcal{C}$ from [23] to construct a defining vector basis.

6.3.4 Nilmonoids

Nilmonoids have a *canonical presentation*: indeed every nilsemigroup N has a smallest generating subset, namely the set $X = N \backslash N^2 = Irr(N)$ of its irreducible elements; every nilmonoid $S = N \cup \{1\}$ is similarly generated. The following results, from [33], apply when S is finite.

The congruence $\mathcal{C}$ has a very large zero class $Z = \pi^{-1}0$, whose complement $\mathbb{F} \backslash Z$ is finite.

In what follows, the compatible well order $\prec$ on $\mathbb{F}$ is the lexicographic order induced by an arbitrary total order $<$ on X. In particular, $x > y$ in X implies $x \prec y$ in $\mathbb{F}$. We denote by ω the greatest element of X under $<$, and r_0 denotes the least element of Z under $\prec$.

A remarkable property of every finite nilmonoid S is the existence of a defining vector basis, even if S is not 0-free. The following result is Lemma 1.4 of [33] and is similar to Lemma XIII.5.1 of [31], a.k.a. Lemma 4.2 of [27].

Lemma 6.3.2 *For each $x < \omega$ there exists $m(x) \in M$ such that $m(x)_x = 1$ and $m(x)_y = 0$ for all $y \neq x, \omega$. Moreover, $r_0 + \omega \in M$.*

Corollary 2 *If S is a finite commutative nilsemigroup, then the vectors $\overrightarrow{m(x)}$ constitute a defining vector basis.*

Corollary 2 is used in [33] to improve the computation of $H^2(S, \mathcal{G})$ in various particular cases: if every element $a \neq 0$ of S whose $\mathcal{C}$-class has more than one element is a minimal element of S; if $\mathcal{G}$ is surjecting; if $\mathcal{G}$ is semiconstant (as defined in Sect. 7.3 below).

6.3.5   Semigroups with Zero Cohomology

This last part addresses two questions:

when is $H^2(S, \mathcal{G}) = 0$ for all $\mathcal{G}$?
if S is finite groupfree, when is $H^2(S, \mathcal{G}) = 0$ for all $\mathcal{G}$ that are thin and surjecting?

The first question was answered as follows in [29] if S is finite groupfree, using the overpath method. Call an abelian group valued functor $\mathcal{G}$ on S *selective* if there exists $c \in S$ such that $G_a = 0$ for all $a \neq c$ and $\gamma_{c,t}$ is the identity on G_c whenever $ct = c$. This makes $\mathcal{G}$ thin but nor surjecting (unless $c = 1$).

Theorem 6.3.2 ([29]) *For a finite groupfree commutative semigroup S the following conditions are equivalent:*

(a) $H^2(S, \mathcal{G}) = 0$ for all $\mathcal{G}$;
(b) $H^2(S, \mathcal{G}) = 0$ for all $\mathcal{G}$ that are selective;
(c) S is a semilattice.

This is Theorem XIV.1.8 of [31].

The second question was answered in [30] but only if S is a finite nilmonoid. A *reaching basis* of $\mathbb{G}$ (called an *optimal basis* in [30] and a *defining basis* in [31]) is a subset B of $M \backslash Z$, where Z is the zero class, such that (i) the defining vectors $\overrightarrow{m}$ with $m \in B$ are distinct and constitute a basis of $\mathbb{G}$ (the free abelian group on X), and (ii) the basic relation $\overrightarrow{m} = \sum_{b \in B} k_{m,b} \overrightarrow{b}$ is reachable in C_m for every $m \in M \backslash B$.

Theorem 6.3.3 ([30]) *For a finite commutative nilsemigroup S the following conditions are equivalent:*

(a) $H^2(S, \mathcal{G}) = 0$ for all $\mathcal{G}$ that are thin and surjecting;
(b) in the canonical presentation of S, $\mathbb{G}$ has a reaching basis;
(c) in every presentation $\pi : F \longrightarrow S$, where F is finitely generated, there is a reaching basis.

This is the main part of Theorem XIV.3.4 of [31].

If S is a finite commutative nilsemigroup, then S has a defining vector basis, by Corollary 2, but there might not be enough elements of $M \backslash Z$ to make a basis of $\mathbb{G}$, let alone a reaching basis; for instance, this is the case if S is 0-free, for then $M \subseteq Z$. Moreover, if $\mathbb{G}$ has a basis that consists of defining vectors from $M \backslash Z$, then each basic relation $\overrightarrow{m} = \sum_{b \in B} k_{m,b} \overrightarrow{b}$ is reachable in some $\mathcal{C}$-class, but might not be reachable in C_m. The following example is from [30].

Example 5   Let S be the commutative nilsemigroup

$$S = \langle a, b \mid a^3 b = b^5, \ a^4 = a^2 b^3, a^5 = a^3 b^2 = a^2 b^5 = b^6 = 0 \rangle.$$

In the canonical presentation of S, the generators of $\mathbb{F}$ are $x = (1, 0)$ and $y = (0, 1)$. Order $\mathbb{F}$ lexicographically so that $(i, j) \prec (k, \ell)$ if and only if either $i < k$, or $i = k$ and $j < \ell$.

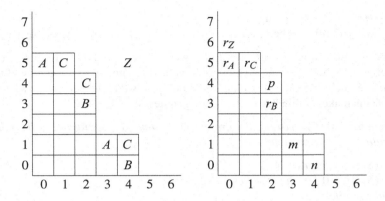

The nontrivial $\mathcal{C}$-class es are

$$A = \{(0, 5), (3, 1)\}, \quad B = \{(2, 3), (4, 0)\}, \quad C = \{1, 5), (2, 4), (4, 1)\},$$

and the zero class Z; their least elements (under $\prec$) are

$$r_A = (0, 5), \quad r_B = (2, 3), \quad r_C = (1, 5), \quad \text{and} \quad r_Z = (0, 6).$$

Hence $M \backslash Z$ has three elements:

$$m = (3, 1), \quad n = (4, 0), \quad \text{and} \quad p = (2, 4),$$

with defining vectors

$$\overrightarrow{m} = (3, -4), \quad \overrightarrow{n} = (2, -3), \quad \text{and} \quad \overrightarrow{p} = (1, -1).$$

We see that $\overrightarrow{n}$ and $\overrightarrow{p}$ constitute a basis of $\mathbb{G}$ (since $\begin{vmatrix} 2 & 1 \\ -3 & -1 \end{vmatrix} = 1$). There is one basic relation $\overrightarrow{m} = \overrightarrow{n} + \overrightarrow{p}$. This relation is reachable in C_p and in Z, but it is not reachable in $C_m = \{(0, 5), (3, 1)\}$, which has only one path $(3, 1) \xrightarrow{m} (0, 5)$. Hence $\{m, n\}$ is not a reaching basis.

On the other hand, $\vec{m}$ and $\vec{n}$ constitute a basis of $\mathbb{G}$ (since $\begin{vmatrix} 3 & 2 \\ -4 & -3 \end{vmatrix} = -1$).

There is one basic relation $\vec{p} = \vec{m} - \vec{n}$. This relation is reachable in $C_p = \{(1, 5), (2, 4), (4, 1)\}$, which has two paths

$$(4, 1) \xrightarrow{m} (1, 5) \text{ and } (4, 1) \xrightarrow{n} (2, 4) \xrightarrow{p} (1, 5).$$

Therefore $B = \{m, n\}$ is a reaching basis, and $H^2(S, \mathcal{G}) = 0$ whenever $\mathcal{G}$ is thin and surjecting. $\qquad\square$

Chapter 7
Symmetric Chains

Symmetry in this monograph refers to certain conditions and to the objects, such as cochains, that satisfy them. This chapter contains general definitions and properties of symmetric sets, mappings, and chains. The latter are then used to construct a projective complex whose cohomology is the symmetric cohomology, and to obtain universal coefficients theorems when the coefficient functor is constant or semiconstant (constant on a subset and null on its complement).

7.1 Symmetric Mappings

This section gives the definition and basic properties of symmetric sets and mappings.

7.1.1 Symmetry

Symmetric mappings are mappings that satisfy certain symmetry conditions. The following definitions are from [34].

A *symmetric set* of order $n \geq 1$ on a set S is a subset X of the cartesian product $S^{(n)} = S \times S \times \cdots \times S$ (with n terms) such that

$$(x_1, x_2, \ldots, x_n) \in X \text{ implies } (x_{\sigma 1}, x_{\sigma 2}, \ldots, x_{\sigma n}) \in X$$

for every permutation σ of $1, 2, \ldots, n$.

A mapping f of a symmetric set X of order $n = 2, 3, 4$ into an abelian group G is *symmetric* if and only if it satisfies the condition (Sn) below. (An appropriate condition (Sn) have not been determined when $n \geq 5$.)

© The Author(s), under exclusive license to Springer Nature Switzerland AG 2022
P. A. Grillet, *The Cohomology of Commutative Semigroups*, Lecture Notes
in Mathematics 2307, https://doi.org/10.1007/978-3-031-08212-2_7

Condition (S2) is

$$f(b, a) = f(a, b), \tag{S2}$$

for all $(a, b) \in X$.

Condition (S3) consists of two conditions:

$$f(c, b, a) = -f(a, b, c), \quad \text{and} \tag{S3a}$$

$$f(a, b, c) = f(b, a, c) - f(b, c, a), \tag{S3b}$$

for all $(a, b, c) \in X$. As in [34], (S3a) follows from (S3b); it is included here to show its analogy to (S2) and (S4a).

Condition (S4) consists of three conditions:

$$f(d, c, b, a) = -f(a, b, c, d), \tag{S4a}$$

$$f(a, b, c, d) = f(b, a, c, d) - f(b, c, a, d) + f(b, c, d, a), \quad \text{and} \tag{S4b}$$

$$f(a, b, b, a) = 0, \tag{S4c}$$

for all $(a, b, c, d) \in X$ (in (S4c), for all $a, b \in S$ such that $(a, b, b, a) \in X$).

Condition (Sn) is basically the same condition that defied symmetric n-cochains in Chap. 3. Indeed, if S is a commutative semigroup and $s \in S$, then the set

$$X_s = \{(a_1, a_2, \ldots, a_n) \mid a_1, a_2, \ldots, a_n \in S, \; a_1 a_2 \cdots a_n = s\}$$

is a symmetric set of order n. An n-cochain $u \in C^n(S, \mathcal{G})$, where $n \le 4$, is symmetric if and only if

$$(a_1, a_2, \ldots, a_n) \longmapsto u(a_1, a_2, \ldots, a_n)$$

is a symmetric mapping of X_s into G_s, for every $s \in S$.

7.1.2 Bases

It is clear from (Sn) that some values of a symmetric mapping are determined by other values. In fact, some values of a symmetric mapping determine all the other values.

A *basis* of a symmetric set X of order $n = 2, 3, 4$ is a subset Y of X such that every mapping of Y into an abelian group G extends uniquely to a symmetric mapping of X into G.

Theorem 7.1.1 *Every symmetric set of order $n = 2, 3, 4$ has a basis.*   □

Theorem 7.1.1 is proved by placing an arbitrary total order on the underlying set S. An explicit basis is then obtained as follows.

Proposition 7.1.1 *Let X be a symmetric set of order $n = 2, 3, 4$ on a totally ordered set S. If $n = 2$, then*

$$Y = \{(a, b) \in X \mid a \le b\}$$

is a basis of X. If $n = 3$, then

$$Y = \{(a, b, c) \in X \mid a \le b \ and \ a < c\}$$

is a basis of X. If $n = 4$, then

$$Y = \{(a, b, c, d) \in X \mid either \ a < b, c, d, \ or \ a \le b, c \ and \ b < d, \ or \ both\},$$

is a basis of X.

Moreover, if f is a mapping of Y into an abelian group G and $\widehat{f}$ is the symmetric mapping of X into G that extends f, then every value of $\widehat{f}$ is a sum of values of f and opposites of values of f. □

The set Y in Proposition 7.1.1 is the *standard basis* of X, relative to the given total order on S.

Proposition 7.1.1 is clear if $n = 2$ and is proved as follows if $n = 3, 4$.

Let $n = 3$. Every triple (x, y, z) of elements of S falls into exactly one of the following cases:

(1) $x = z$,

(2) $x = y < z$,

(3) $x = y > z$,

(4) $x < y = z$,

(5) $x > y = z$,

(6) $x < y < z$ or $x < z < y$,

(7) $y < x < z$ or $y < z < x$,

(8) $z < x < y$ or $z < y < x$.

Lemma 7.1.1 *If S is totally ordered and $n = 3$, then*
(1) *a mapping $f : X \longrightarrow G$ is symmetric if and only if, for all $(x, y, z) \in X$,*

$$if \ x = z, \ then \ f(x, y, z) = 0, \tag{P1}$$

$$if \ x = y > z, \ then \ f(x, y, z) = -f(z, y, x), \tag{P3}$$

$$\text{if } x > y = z, \text{ then } f(x, y, z) = -f(z, y, x), \tag{P5}$$

if $y < x < z$ *or if* $y < z < x$, *then*

$$f(x, y, z) = f(y, x, z) - f(y, z, x), \tag{P7}$$

if $z < x < y$ *or if* $z < y < x$, *then* $f(x, y, z) = -f(z, y, x). \tag{P8}$

Moreover, (2) *the set* $Y = \{(a, b, c) \in X \mid a \le b \text{ and } a < c\}$ *contains every* (a, b, c) *that appears in the right hand side of* (P1), ..., (P8).

Proof If f is symmetric, then (P1) and (P7) follows from (S3b), and (P3), (P5), (P8) follow from (S3a). Moreover, the term(s) (a, b, c) that appear in the right hand sides of (P3), (P5), (P7), and (P8) are in cases (4), (2), (6), and (6), respectively, which puts them in Y.

Conversely, assume that f has properties (P1) through (P8).

If $a = c$, then $f(b, a, c) - f(b, c, a) = 0 = f(a, b, c)$, by (P1), and (S3b) holds.

If $a = b < c$, or if $a = b > c$, then $f(b, c, a) = 0$ by (P1), $f(b, a, c) = f(a, b, c)$, and (S3b) holds.

If $a < b = c$, then $f(b, a, c) = 0$ by (P1), $f(b, c, a) = -f(a, c, b) = -f(a, b, c)$ by (P3), and (S3b) holds.

If $a > b = c$, then $f(b, a, c) = 0$ by (P1), $f(a, b, c) = -f(c, b, a) = -f(b, c, a)$ by (P5), and (S3b) holds.

If $a < b < c$, or if $a < c < b$, then $f(b, a, c) = f(a, b, c) - f(a, c, b)$ by (P7), $f(b, c, a) = -f(a, c, b)$ by (P8), and (S3b) holds.

If $b < a < c$, or if $b < c < a$, then $f(a, b, c) = f(b, a, c) - f(b, c, a)$ by (P7), and (S3b) holds.

If $c < a < b$, or if $c < b < a$, then $f(b, a, c) = -f(c, a, b)$ by (P8), $f(b, c, a) = f(c, b, a) - f(c, a, b)$ by (P7), $f(a, b, c) = -f(c, b, a)$ by (P8), and (S3b) holds.

In every case (S3b) holds; therefore f is symmetric.   □

We can now prove Proposition 7.1.1 in case $n = 3$. Let f be a mapping of Y into an abelian group G. By part (2) of Lemma 7.1.1, a mapping $\widehat{f} \colon X \longrightarrow G$ may be defined by

$$\widehat{f}(a, b, c) = \begin{cases} (1) & 0 & \text{if } a = c, \\ (2) & f(a, b, c) & \text{if } a = b < c, \\ (3) & -f(c, b, a) & \text{if } a = b > c, \\ (4) & f(a, b, c) & \text{if } a < b = c, \\ (5) & -f(b, c, a) & \text{if } a > b = c, \\ (6) & f(a, b, c) & \text{if } a < b < c \text{ or if } a < c < b, \\ (7) & f(b, a, c) - f(b, c, a) & \text{if } b < a < c \text{ or if } b < c < a, \\ (8) & -f(c, b, a) & \text{if } c < a < b \text{ or if } c < b < a, \end{cases}$$

for all $(a, b, c) \in X$. In fact, if $\widehat{f}$ is to extend f and be symmetric, then $\widehat{f}$ must have properties (P1), ..., (P8) and be defined as above. Conversely, $\widehat{f}$, as defined, extends f and has properties (P1), ..., (P8); hence, by Lemma 7.1.1, $\widehat{f}$ is symmetric.

The proof of Proposition 7.1.1 in case $n = 4$ is similar and equally straightforward. However, 4-tuples (w, x, y, z) now fall into 56 different cases. This causes a much longer proof, which fills most of Section 7 of [34].

7.2 Chain Groups

This section contains the definition and basic properties of symmetric chains and chain groups, after [35].

7.2.1 Definition

Let X be a symmetric set of order $n = 1, 2, 3, 4$ on a set S. The *chain group* $C(X)$ of X is the abelian group generated by all $\langle a_1, a_2, \ldots, a_n \rangle$ such that $(a_1, a_2, \ldots, a_n) \in X$, subject to the condition (Cn) defined below. A *symmetric chain* of X is an element of $C(X)$.

Condition (C1) is vacuous. If $n = 1$, then $C(X)$ us, up to isomorphism, the free abelian group on X.

Condition (C2) applies if $n = 2$:

$$\langle b, a \rangle = \langle a, b \rangle, \tag{C2}$$

for all $(a, b) \in X$.

Condition (C3) applies if $n = 3$ and consists of two conditions:

$$\langle c, b, a \rangle = -\langle a, b, c \rangle, \quad \text{and} \tag{C3a}$$

$$\langle a, b, c \rangle = \langle b, a, c \rangle - \langle b, c, a \rangle, \tag{C3b}$$

for all $(a, b, c) \in X$. As in Sect. 7.1, (C3a) follows from (C3b); it is included here so that it would not be missed.

Condition (C4) applies if $n = 4$ and consists of three conditions:

$$\langle d, c, b, a \rangle = -\langle a, b, c, d \rangle, \tag{C4a}$$

$$\langle a, b, c, d \rangle = \langle b, a, c, d \rangle - \langle b, c, a, d \rangle + \langle b, c, d, a \rangle, \quad \text{and} \tag{C4b}$$

$$\langle a, b, b, a \rangle = 0, \tag{C4c}$$

for all $(a, b, c, d) \in X$ (in (C4c), for all $a, b \in S$ such that $(a, b, b, a) \in X$).

7.2.2   Properties

The group $C(X)$ comes with a mapping $\iota : X \longrightarrow C(X)$ that sends $(a_1, \ldots, a_n) \in X$ to $\langle a_1, \ldots, a_n \rangle \in C(X)$. The defining relations of $C(X)$ in effect state that ι is a symmetric mapping of X into $C(X)$.

The presentation of $C(X)$ by generators and relations implies that ι is a universal symmetric mapping:

Proposition 7.2.1 *Every symmetric mapping f of X into an abelian group G extends uniquely (via ι) to a homomorphism of $C(X)$ into G: there is a unique homomorphism $\varphi : C(X) \longrightarrow G$ such that $\pi \circ \iota = f$.*   □

Combining Proposition 7.2.1 and Theorem 7.1.1 yields

Theorem 7.2.1 *If X is a symmetric set of order $n = 1, 2, 3, 4$, then $C(X)$ is a free abelian group. Moreover, if Y is a basis of X, then $\iota(Y)$ is a basis of $C(X)$, and ι is injective on Y (if $x, y \in Y$ and $\iota(x) = \iota(y)$, then $x = y$).*

Proof Every mapping of Y into an abelian group G extends uniquely to X and thence extends uniquely (via ι) to $C(X)$.   □

The next result (not in [34] or [35]) shows that $C(X)$ and ι are faithfully inherited by subsets.

Proposition 7.2.2 *Let X and X' be symmetric sets of order $n \leq 4$ on the same set S. If $X \subseteq X'$, then: there is a unique homomorphism $\kappa : C(X) \longrightarrow C(X')$ such that $\kappa \circ \iota$ and ι' agree on X; κ is injective; if Y and Y' are the standard bases of X and X' relative to the same total order on S, then $Y = Y' \cap X$.*

Proof The restriction of ι' to X is a symmetric mapping of X into $C(X')$ and must by Proposition 7.2.1 factor uniquely through ι. Thus κ sends $\langle a_1, \ldots, a_n \rangle \in C(X)$ to $\langle a_1, \ldots, a_n \rangle \in C(X')$, for every $(a_1, \ldots, a_n) \in X$.

The last part of the statement, $Y = Y' \cap X$, follows from the definition of standard bases: if $a = (a_1, \ldots, a_n) \in X$, then a satisfies the inequalities in Proposition 7.1.1 as an element of X if and only if it satisfies the same inequalities as an element of X'. In particular, $Y \subseteq Y'$.

Finally, $C(X)$ is (in particular) generated by $\iota(Y)$; hence $\kappa(C(X))$ is generated by $\kappa(\iota(Y)) = \iota'(Y)$, which is a subset of the basis $\iota'(Y')$ of $C(X')$. Therefore $\kappa(C(X))$ is a free abelian group and $\kappa(\iota(Y))$ is a basis of $\kappa(C(X))$. Moreover, κ is injective on $\iota(Y)$, since $\kappa \circ \iota = \iota'$ is injective on Y'. Hence κ is an isomorphism of $C(X)$ onto $\kappa(C(X))$.   □

In what follows we take advantage of Proposition 7.2.2 and regard κ as an inclusion homomorphism, so that $\langle a_1, \ldots, a_n \rangle$ is the same in $C(X)$ and $C(X')$ when $(a_1, \ldots, a_n) \in X \subseteq X'$. For future use we restate Proposition 7.2.2 as follows:

Proposition 7.2.3 *Let X and X' be symmetric sets of order $n \leq 4$ on the same set S. If $X \subseteq X'$, then $C(X) \subseteq C(X')$; the canonical mappings $\iota: X \longrightarrow C(X)$ and $\iota': X' \longrightarrow C(X')$ agree on X; if Y and Y' are the standard bases of X and X' relative to the same total order on S, then $Y = Y' \cap X$.* $\qquad\square$

7.2.3 Symmetric n-chains

A *symmetric n-chain* of a commutative semigroup S, where $n = 1, 2, 3, 4$, is an element of $C_n(S) = C(S^{(n)})$.

As above, $C_n(S)$ is the abelian group generated by all $\langle a_1, a_2, \ldots, a_n \rangle$ with $a_1, a_2, \ldots, a_n \in S$, subject to (Cn).

The group $C_n(S)$ comes with a canonical mapping

$$\iota: S^{(n)} \longrightarrow C_n(S), \quad (a_1, a_2, \ldots, a_n) \longmapsto \langle a_1, a_2, \ldots, a_n \rangle.$$

By Proposition 7.2.1, every symmetric mapping of $S^{(n)}$ into an abelian group G extends uniquely (via ι) to a homomorphism of $C_n(S)$ into G. Moreover, Theorem 7.2.1 yields:

Theorem 7.2.2 *If S is a commutative semigroup and $n = 1, 2, 3, 4$, then $C_n(S)$ is a free abelian group. Moreover, if Y is a basis of $S^{(n)}$, then $\iota(Y)$ is a basis of $C_n(S)$ and ι is injective on Y.*

The group $C_n(S)$ can be analyzed as follows. For each $s \in S$ we saw that

$$X_s = \{(a_1, a_2, \ldots, a_n) \mid a_1, a_2, \ldots, a_n \in S, \ a_1 a_2 \cdots a_n = s\} \subseteq S^{(n)}$$

is a symmetric set of order n, and $S^{(n)}$ is the disjoint union $S^{(n)} = \bigcup_{s \in S} X_s$ of all X_s.

Proposition 7.2.4 *If $n = 1, 2, 3, 4$, then*

$$C_n(S) = \bigoplus_{s \in S} C(X_s).$$

Relative to any total order on S, the standard basis Y of $S^{(n)}$ is the disjoint union $Y = \bigcup_{s \in S} Y_s$ of the standard bases Y_s of all X_s.

Proof By Proposition 7.2.3, $C(X_s)$ is a subgroup of $C_n(S)$, and $Y_s = Y \cap X_s$. Hence $Y = \bigcup_{s \in S} Y_s$. Since ι is injective on Y, the basis $\iota(Y)$ of $C_n(S)$ is the disjoint union $\iota(Y) = \bigcup_{s \in S} \iota(Y_s)$ of the bases $\iota(Y_s)$ of all $C(X_s)$. Therefore $C_n(S) = \bigoplus_{s \in S} C(X_s)$. $\qquad\square$

The groups $C_n(S)$ come with *boundary homomorphism s*:

Proposition 7.2.5 *If $n = 2, 3, 4$, then there exists a unique homomorphism*

$$\partial : C_n(S) \longrightarrow C_{n-1}(S)$$

such that

$$\partial\langle a, b\rangle = \langle b\rangle - \langle ab\rangle + \langle a\rangle \qquad \text{if } n = 2,$$

$$\partial\langle a, b, c\rangle = \langle b, c\rangle - \langle ab, c\rangle + \langle a, bc\rangle - \langle a, b\rangle \qquad \text{if } n = 3,$$

$$\partial\langle a, b, c, d\rangle = \langle b, c, d\rangle - \langle ab, c, d\rangle$$
$$+ \langle a, bc, d\rangle - \langle a, b, cd\rangle + \langle a, b, c\rangle \qquad \text{if } n = 4,$$

for all $a, b, c, d \in S$. Moreover, $\partial \circ \partial = 0$.

Proof The mapping $f : S \times S \longrightarrow C_1(S)$ defined by:

$$f(a, b) = \langle b\rangle - \langle ab\rangle + \langle a\rangle$$

for all $a, b \in S$, has property (C2): $f(b, a) = f(a, b)$, for all $a, b \in S$. By Proposition 7.2.4, $C_2(S)$ is generated by all $\langle a, b\rangle$ subject to all relations (C2). Therefore f induces a unique homomorphism $\partial : C_2(S) \longrightarrow C_1(S)$ such that $\partial\langle a, b\rangle = f(a, b)$ for all $a, b \in S$. The proof in case $n = 3$ or $n = 4$ is equally straightforward, just longer. Finally,

$$\partial\partial\langle a, b, c\rangle = \partial\langle b, c\rangle - \partial\langle ab, c\rangle + \partial\langle a, bc\rangle - \partial\langle a, b\rangle$$

$$= \langle c\rangle - \langle bc\rangle + \langle b\rangle - \langle c\rangle + \langle abc\rangle - \langle ab\rangle$$

$$+ \langle bc\rangle - \langle abc\rangle + \langle a\rangle - \langle b\rangle + \langle ab\rangle - \langle a\rangle$$

$$= 0$$

for all $a, b, c \in S$, whence $\partial\partial c = 0$ for all $c \in C_3(S)$. (Other considerations complicate the similar calculation in [34].) That $\partial\partial c = 0$ for all $c \in C_4(S)$ is proved similarly.   □

For the sake of completeness, $\partial : C_1(S) \longrightarrow C_0(S) = 0$ is 0.

A *symmetric n-cycle* of S, where $n = 1, 2, 3, 4$, is a symmetric n-chain $c \in C_n(S)$ such that $\partial c = 0$; a *symmetric n-boundary* of S, where $n = 1, 2, 3$, is a symmetric n-chain $b \in C_n(S)$ such that $b = \partial c$ for some $c \in C_{n+1}(S)$. Under pointwise addition, symmetric n-cycles and boundaries constitute subgroups $B_n(S) \subseteq Z_n(S)$ of $C_n(S)$ ($n = 1, 2, 3$). The *n-th symmetric homology group $H_n(S)$* of S, where $n = 1, 2, 3$, is

$$H_n(S) = Z_n(S) / B_n(S).$$

By Prop. 2.5 of [34], $H_1(S)$ is the universal abelian group of S. On the other hand, $H_n(S) = 0$ for all $n = 1, 2, 3$ if S has a zero element (Prop. 2.6 of [34]).

7.3   Chain Functors

As we shall see in the next section, the symmetric homology groups $C_n(S)$ and their cousins are glad to be of service to the Eilenberg-MacLane style cohomologies of commutative semigroups. But general commutative semigroup cohomology is happier with symmetric chain functors.

In this section we use chain groups to construct complexes of chain functors, of which H_S^n is the cohomology. This requires functors $\mathcal{F}_n$ such that $C_S^n(S, \mathcal{G}) \cong \mathrm{Hom}\,(\mathcal{F}_n, \mathcal{G})$. Necessarily $\mathcal{F}_n$ depends on the category we use for $\mathcal{F}_n$ and $\mathcal{G}$, since $\mathrm{Hom}\,(\mathcal{F}_n, \mathcal{G})$ depends on that category but $C_S^n(S, \mathcal{G})$ does not. In this section we use the category $\mathcal{T}$ of thin abelian group valued functors on S and the category $\mathcal{A}$ of *all* abelian group valued functors on S. A third category is considered in the next section.

The Beck cohomology of S is already the cohomology of a projective chain complex, which would serve here ([31], Prop. XII.4.8). Overweight cochains, however, beget overweight chains, and the functors constructed here, though limited to small dimensions, use more natural and convenient chains.

7.3.1   *Thin Chain Functors*

Given a commutative semigroup S and $n = 1, 2, 3, 4$, construct an abelian group valued functor $\mathcal{C}_n(S)$ on S as follows. For each $a \in S$,

$$X_a^\uparrow = \{(a_1, a_2, \ldots, a_n) \mid a_1, a_2, \ldots, a_n \in S,\ a_1 a_2 \cdots a_n \geq a\} \subseteq S^{(n)}$$

is a symmetric set. Let

$$C_n(a) = C(X_a^\uparrow) \subseteq C(S^{(n)}) = C(S).$$

Thus $C_n(a)$ is the abelian group generated by all $\langle a_1, a_2, \ldots, a_n \rangle$ such that $a_1, a_2, \ldots, a_n \in S$ and $a_1 a_2 \cdots a_n \geq a$, subject to (C$n$). By Theorem 7.2.1, $C_n(a)$ is a free abelian group. By Proposition 7.2.3, $C_n(a)$ is also the subgroup of $C(S)$ generated by all $\langle a_1, a_2, \ldots, a_n \rangle$ such that $a_1 a_2 \cdots a_n \geq a$, and (since $X_a^\uparrow = \bigcup_{s \geq a} X_s$) the direct sum

$$C_n(a) = \bigoplus_{s \geq a} C(X_s).$$

If $a \geq b$, then $X_a^\uparrow \subseteq X_b^\uparrow$ and $C_n(a) \subseteq C_n(b)$, by Proposition 7.2.3.

By definition, $\mathcal{C}_n(S)$ is the thin abelian group valued functor that assigns $C_n(a)$ to each $a \in S$ and the inclusion homomorphism $\gamma_b^a : C_n(a) \longrightarrow C_n(b)$ to each $a, b \in S$ such that $a \geq b$.

The main properties of $\mathcal{C}_n(S)$ are as follows.

Proposition 7.3.1 *If $n = 1, 2, 3, 4$, then $\mathcal{C}_n(S)$ is projective in the category $\mathcal{T}$ of thin abelian group valued functors on S.* □

This is Theorem 4.4 of [34]. The proof uses Theorem 7.1.1 to show that $C_S^n(S, -)$ preserves epimorphisms. Then Proposition 7.3.1 follows from Proposition 7.3.2 below.

The universal property of $C(X_s)$ yields the following property of $\mathcal{C}_n(S)$ (part of Prop. 4.2 of [34]):

Proposition 7.3.2 *Let $n = 1, 2, 3, 4$ and let $\mathcal{G} = (G, \gamma)$ is a thin abelian group valued functor on S. There is an isomorphism*

$$U : \mathrm{Hom}_{\mathcal{T}}(\mathcal{C}_n(S), \mathcal{G}) \cong C_S^n(S, \mathcal{G})$$

which is natural in $\mathcal{G}$ and sends a natural transformation $\tau : \mathcal{C}_n(S) \longrightarrow \mathcal{G}$ to the symmetric n-cochain u defined by

$$u(a_1, a_2, \ldots, a_n) = \tau_a \langle a_1, a_2, \ldots, a_n \rangle, \quad \text{where } a = a_1 a_2 \cdots a_n.$$

Then $\tau_a \langle a_1, \ldots, a_n \rangle = \gamma_a^s u(a_1, \ldots, a_n)$ whenever $s = a_1 \cdots a_n \geq a$. □

Finally, the homomorphism $\partial : C_n(S) \longrightarrow C_{n-1}(S)$ sends $C_n(a)$ into $C_{n-1}(a)$. Hence $\mathcal{C}_n(S)$ inherits boundaries from $C_n(S)$:

Proposition 7.3.3 *If $n = 1, 2, 3, 4$, then $\partial : C_n(S) \longrightarrow C_{n-1}(S)$ induces a natural transformation $\partial : \mathcal{C}_n(S) \longrightarrow \mathcal{C}_{n-1}(S)$. Moreover, the following square commutes:*

$$
\begin{array}{ccc}
\mathrm{Hom}_{\mathcal{T}}(\mathcal{C}_{n-1}(S), \mathcal{G}) & \xrightarrow{\partial^*} & \mathrm{Hom}_{\mathcal{T}}(\mathcal{C}_n(S), \mathcal{G}) \\
\downarrow{\scriptstyle U} & & \downarrow{\scriptstyle U} \\
C_S^{n-1}(S, \mathcal{G}) & \xrightarrow{\delta} & C_S^n(S, \mathcal{G})
\end{array}
$$

where U is the isomorphism in Proposition 7.3.2 and $\partial^* \tau = \tau \circ \partial$. □

We now have a short projective complex

$$\mathcal{C}_*(S) : 0 \longleftarrow \mathcal{C}_1(S) \xleftarrow{\partial} \cdots \xleftarrow{\partial} \mathcal{C}_4(S)$$

of thin abelian group valued functors, whose cohomology with coefficients in $\mathcal{G}$ is the symmetric cohomology of S if $\mathcal{G}$ is thin:

Theorem 7.3.1 *If $n = 1, 2, 3$ and $\mathcal{G}$ is thin, then there is an isomorphism*

$$H_S^n(S, \mathcal{G}) \cong H^n(\mathcal{C}_*(S), \mathcal{G})$$

which is natural in $\mathcal{G}$.

7.3.2 General Chain Functors

Proposition 7.3.2 determines $\mathcal{C}_n(S)$ up to isomorphism. Removing the thin requirement from this result and from Theorem 7.3.1 results in a different functor, as expected.

If $a_1 a_2 \cdots a_n \geq a$, then $a_1 a_2 \cdots a_n t = a$ for some $t \in S^1$. As in Section 4 of [34], the new functor $A_n(S)$ is constructed by adding t to the variables $a_1, a_2, \ldots, a_n$; this spreads copies of the elements of $C_n(a)$ among the various values of t. In detail:

Let $A_0(a) = 0$ for all $a \in S$.

Let $A_1(a)$ be the free abelian group on the set

$$\{\langle x; t \rangle \mid x \in S, \ t \in S^1, \ xt = a\}.$$

Let $A_2(a)$ be the abelian group generated by all $\langle x, y; t \rangle$ such that $x, y \in S$, $t \in S^1$, and $xyt = a$, subject to all defining relations

$$\langle y, x; t \rangle = \langle x, y; t \rangle. \tag{C2}$$

Let $A_3(a)$ be the abelian group generated by all $\langle x, y, z; t \rangle$ such that $x, y, z \in S$, $t \in S^1$, and $xyzt = a$, subject to all defining relations

$$\langle z, y, x; t \rangle = -\langle x, y, z; t \rangle, \ \text{and} \tag{C3a}$$

$$\langle x, y, z; t \rangle = \langle y, x, z; t \rangle - \langle y, z, x; t \rangle. \tag{C3b}$$

Let $A_4(a)$ be the abelian group generated by all $\langle w, x, y, z; t \rangle$ such that $w, x, y, z \in S, t \in S^1$, and $wxyzt = a$, subject to all defining relations

$$\langle z, y, x, w; t \rangle = -\langle w, x, y, z; t \rangle, \tag{C4a}$$

$$\langle w, x, y, z; t \rangle = \langle x, w, y, z; t \rangle - \langle x, y, w, z; t \rangle + \langle x, y, z, w; t \rangle, \ \text{and} \tag{C4b}$$

$$\langle x, y, y, x; t \rangle = 0. \tag{C4c}$$

The groups $A_n(a)$ can be analyzed as follows. For each $s \in S$ and $t \in S^1$ let $A_n(s; t)$ be the subgroup of $A_n(st)$ generated by all $\langle x_1, x_2, \ldots, x_n; t \rangle$ such that $x_1 x_2 \cdots x_n = s$.

Lemma 7.3.1 *Let $n = 1, 2, 3, 4$, $s \in S$, and $t \in S^1$. The mapping*

$$\langle x_1, x_2, \ldots, x_n \rangle \longmapsto \langle x_1, x_2, \ldots, x_n; t \rangle,$$

where $x_1 x_2 \cdots x_n = s$, induces an isomorphism of $C(X_s)$ onto $A_n(s; t)$. Hence $A_n(s; t)$ is a free abelian group. Moreover, $A_n(a) \cong \bigoplus_{st=a} A_n(s; t)$. Hence $A_n(a)$ is a free abelian group. $\square$

This is Lemma 5.1 of [34]. It implies universal properties that yield similar properties for $A_n(a)$ and $C_n(a)$.

Next, for each $n = 1, 2, 3, 4$, $a \in S$, and $u \in S^1$, there is a unique homomorphism $\alpha_{a,u} : A_n(a) \longrightarrow A_n(au)$ such that

$$\alpha_{a,u} \langle x_1, \ldots, x_n; t \rangle = \langle x_1, \ldots, x_n; tu \rangle$$

whenever $x_1 \cdots x_n t = a$. Moreover, $\alpha_{a,1}$ is the identity on $A_n(a)$ and $\alpha_{au,v} \circ \alpha_{a,u} = \alpha_{a,uv}$, for all a, u, v.

By definition, $\mathcal{A}_n(S) = (A_n, \alpha)$ is the abelian group valued functor on S that assigns $A_n(a)$ to each $a \in S$ and $\alpha_{a,u} : A_n(a) \longrightarrow A_n(au)$ to each $a \in S$ and $u \in S^1$.

The main properties of $\mathcal{A}_n(S)$ are as follows.

Proposition 7.3.4 ([34], Theorem 5.5) *If $n = 1, 2, 3, 4$, then $\mathcal{A}_n(S)$ is projective in the category $\mathcal{A}$ of [all] abelian group valued functors on S.* $\square$

Proposition 7.3.5 ([34], Prop. 5.4) *If $n = 1, 2, 3, 4$ and $\mathcal{G} = (G, \gamma)$ is an abelian group valued functor on S, then there is an isomorphism*

$$U : \mathrm{Hom}_{\mathcal{A}} (\mathcal{A}_n(S), \mathcal{G}) \cong C_S^n(S, \mathcal{G})$$

which is natural in $\mathcal{G}$ and sends a natural transformation $\tau : \mathcal{A}_n(S) \longrightarrow \mathcal{G}$ to the symmetric n-cochain u defined by

$$u(x_1, \ldots, x_n) = \tau_a \langle x_1, \ldots, x_n; 1 \rangle, \quad \text{where } a = x_1 \cdots x_n.$$

Then $\tau_a \langle x_1, \ldots, x_n; t \rangle = \gamma_{s,t} u(x_1, \ldots, x_n)$ whenever $s = x_1 \cdots x_n$ and $st = a$. $\square$

Lemma 7.3.2 ([34], Lemmas 5.6 and 5.7) *If $n = 1, 2, 3, 4$, then there is a unique natural transformation $\partial : \mathcal{A}_n(S) \longrightarrow \mathcal{A}_{n-1}(S)$ such that*

$$\partial \langle x; t \rangle = 0,$$

$$\partial \langle x, y; t \rangle = \langle y; xt \rangle - \langle xy; t \rangle + \langle x; yt \rangle,$$

$$\partial \langle x, y, z; t \rangle = \langle y, z; xt \rangle - \langle xy, z; t \rangle + \langle x, yz; t \rangle - \langle x, y; zt \rangle,$$

$$\partial \langle w, x, y, z; t \rangle = \langle x, y, z; wt \rangle - \langle wx, y, z; t \rangle$$

$$+ \langle w, xy, z; t \rangle - \langle w, x, yz; t \rangle + \langle w, x, y; zt \rangle,$$

for all $w, x, y, z \in S$ and $t \in S^1$. Moreover, $\partial \circ \partial = 0$, and the following square commutes:

$$
\begin{array}{ccc}
\mathrm{Hom}_{\mathcal{A}}\,(\mathcal{A}_{n-1}(S), \mathcal{G}) & \xrightarrow{\ \partial^* \ } & \mathrm{Hom}_{\mathcal{A}}\,(\mathcal{A}_n(S), \mathcal{G}) \\
{\scriptstyle U}\downarrow & & \downarrow{\scriptstyle U} \\
C_S^{n-1}(S, \mathcal{G}) & \xrightarrow{\ \delta \ } & C_S^n(S, \mathcal{G})
\end{array}
$$

where U is the isomorphism in Proposition 7.3.5 and $\partial^* \tau = \tau \circ \partial$. □

We now have a short projective complex

$$\mathcal{A}_*(S) : 0 \longleftarrow \mathcal{A}_1(S) \xleftarrow{\ \partial \ } \cdots \xleftarrow{\ \partial \ } \mathcal{A}_4(S)$$

of abelian group valued functors, whose cohomology with coefficients in any abelian group valued functor $\mathcal{G}$ on S is the symmetric cohomology of S:

Theorem 7.3.2 *If $n = 1, 2, 3$ and $\mathcal{G}$ is any abelian group valued functor $\mathcal{G}$ on S, then there is an isomorphism*

$$H_S^n(S, \mathcal{G}) \cong H^n(\mathcal{A}_*(S), \mathcal{G})$$

which is natural in $\mathcal{G}$.

Theorem 7.3.2 provides a direct proof of the long exact sequence in Theorem 3.3.2 in Chap. 4.

7.4 Semiconstant Functors

This section studies symmetric cohomology with semiconstant coefficients.

7.4.1 Definition

The definition of semiconstant functors is based on:

Lemma 7.4.1 ([34], Lemma 1.1) *Given a commutative semigroup S, a subset B of S, and an abelian group $G \neq 0$, let*

$$G_a = \begin{cases} G & \text{if } a \in B, \\ 0 & \text{if } a \notin B, \end{cases}$$

for all $a \in S$, and

$$\gamma_{a,t} = \begin{cases} 1_G & \text{if } a \in B \text{ and } at \in B, \\ 0 & \text{if } a \notin B \text{ or } at \notin B, \end{cases}$$

for all $a \in S$ and $t \in S^1$. Then $\mathcal{F}(G/B) = \mathcal{G} = (G, \gamma)$ is an abelian group valued functor on S if and only if B has the following property: for all $a \in S$ and $t, u \in S^1$, if $a \in B$ and $atu \in B$, then $at \in B$; and then $\mathcal{G}$ is thin.

Proof If $a \in B$, $atu \in B$, and $at \notin B$, then $\gamma_{a,tu} = 1_G \neq 0 = \gamma_{at,u} \circ \gamma_{a,t}$ and $\mathcal{G}$ is not a functor. On the other hand, if $a \in B$ and $atu \in B$ imply $at \in B$, then

$$\gamma_{at,u} \circ \gamma_{a,t} = \begin{cases} 0 = \gamma_{a,tu} & \text{if } a \notin B \text{ or if } atu \notin B, \\ 1_G = \gamma_{a,tu} & \text{if } a \in B \text{ and } atu \in B, \end{cases}$$

for then $at \in B$.   □

A subset B of a commutative semigroup S is *convex* if and only if, for all $a \in S$ and $t, u \in S^1$, if $a \in B$ and $atu \in B$, then $at \in B$; equivalently, using the divisibility preorder $\leq$ on S, if $a, c \in B$ and $a \geq b \geq c$, then $b \in B$.

Examples of convex subsets include: all of S; any ideal I of S; the complement $S \backslash I$ of any ideal of S; in particular, if S has a zero element, $S \backslash \{0\}$; the difference $I \backslash J$ between any two ideals I, J of S; and any one-element subset $\{c\}$ of S.

An abelian group valued functor $\mathcal{G}$ on S is *semiconstant* if and only if $\mathcal{G} = \mathcal{F}(G/B)$, as constructed in Lemma 7.4.1, for some abelian group G and convex subset B of S; then $\mathcal{G}$ is semiconstant *on B at G*.

Examples of semiconstant functors include *constant* functors $\mathcal{F}(G, S)$, constant on all of S; if S has a zero element, *almost constant* functors $\mathcal{F}(G, S \backslash \{0\})$, constant on $S \backslash \{0\}$; and *selective* functors $\mathcal{F}(G, \{c\})$ from Sect. 6.2, constant on a one-element subset $\{c\}$ of S.

Cohomology with a constant coefficient functor classifies commutative Rédei extensions (as in [47]; see Sect. 9.1); it is a commutative analogue of the Eilenberg-MacLane cohomology of monoids (see, for instance, [41]). If S has a zero element, then this cohomology vanishes (Proposition 7.4.6 below), and cohomology with an almost constant functor is a better choice; when applied to nilmonoids, it classifies homogeneous elementary semigroups [18].

7.4.2 Chain Groups

The chain groups $C_n(S)$ in Sect. 7.2 adapt as follows for use with semiconstant coefficients.

Let $C_0(S/B) = 0$. Let $n = 1, 2, 3, 4$. The set

$$X_B = \{(a_1, a_2, \ldots, a_n) \mid a_1, a_2, \ldots, a_n \in S, \ a_1 a_2 \cdots a_n \in B\}$$

is a symmetric set on S of order n; then $C_n(S/B)$ is the chain group

$$C_n(S/B) = C(X_B)$$

of X_B: the abelian group generated by a copy

$$\{\langle a_1, a_2, \ldots, a_n \rangle \mid a_1, a_2, \ldots, a_n \in S, \ a_1 a_2 \cdots a_n \in B\}$$

of X_B, subject to all defining relations (Cn). If $n = 1$, then (C1) is vacuous and $C_1(S/B)$ is free on a copy $\{\langle b \rangle \mid b \in B\}$ of B. In general, $C_n(S/B)$ comes with a canonical mapping $\iota \colon X_B \longrightarrow C_n(S/B)$ that sends $(a_1, a_2, \ldots, a_n) \in X_B$ to $\langle a_1, a_2, \ldots, a_n \rangle \in C_n(S/B)$.

For later use it is convenient to define $\langle a_1, a_2, \ldots, a_n \rangle$ for all $a_1, a_2, \ldots, a_n$ (where $n \leq 4$): let

$$\langle a_1, a_2, \ldots, a_n \rangle = 0 \in G_{a_1 a_2 \cdots a_n} = 0 \text{ if } a_1 a_2 \cdots a_n \notin B.$$

Some of the generality implied by the arbitrary convex subset B disappears if $S \backslash B$ is an ideal:

Proposition 7.4.1 ([34], Prop. 2.1) *If $n = 1, 2, 3, 4$ and $B = S \backslash I$, where I is an ideal of S, then*

$$C_n(S/B) = C_n(T, T \backslash \{0\}),$$

where $T = S/I$. □

7.4.3 Properties

The main properties of $C_n(S/B)$ follow from the following analysis. For each $s \in S$, recall that

$$X_s = \{(a_1, a_2, \ldots, a_n) \mid a_1, a_2, \ldots, a_n \in S, \; a_1 a_2 \cdots a_n = s\}.$$

Then X_B is a disjoint union $X_B = \bigcup_{s \in B} X_s$.

Proposition 7.4.2 *If $n = 1, 2, 3, 4$, then*

$$C_n(S/B) \cong \bigoplus_{s \in B} C(X_s) \subseteq C_n(S).$$

Hence $C_n(S/B)$ is a free abelian group; if Y_s is a basis of X_s, then $\iota\left(\bigcup_{s \in B} Y_s\right)$ is a basis of $C_n(S/B)$.   □

This is proved like Proposition 7.2.4. In particular, $C_n(S/S) \cong C_n(S)$. The elements of $C_n(S/B)$ are the *symmetric n-chains of S relative to B*.

Proposition 7.4.3 ([34], Prop. 2.3) *If $n = 1, 2, 3, 4$ and $\mathcal{G} = \mathcal{F}(G/B)$ is semiconstant on B at G, then there is an isomorphism*

$$U : \mathrm{Hom}\,(C_n(S/B), G) \cong C_S^n(S, \mathcal{G})$$

which is natural in $\mathcal{G}$ and sends a homomorphism $\varphi \colon C_n(S/B) \longrightarrow G$ to the cochain $u = U(\varphi)$ defined by

$$u\,(a_1, a_2, \ldots, a_n) = \begin{cases} \varphi\,\langle a_1, a_2, \ldots, a_n \rangle \in G & \text{if } a \in B, \\ 0 \in G_a & \text{if } a \notin B, \end{cases}$$

for all $a_1, a_2, \ldots, a_n \in S$, where $a = a_1 a_2 \cdots a_n$.

Proof This follows from Proposition 7.2.1. Since $G_a = 0$ whenever $a \notin B$, a cochain $u \in C_S^n(S, \mathcal{G})$ is uniquely determined by its values on X_n, which constitute a symmetric mapping that extends uniquely to a homomorphism of $C_n(S/B) = C(X_n)$ into G. Conversely, if $\varphi \colon C_n(S/B) = C(X_n) \longrightarrow G$ is a homomorphism, then $\varphi \circ \iota \colon X_n \longrightarrow G$ is a symmetric mapping and extends uniquely to a cochain $u \in C_S^n(S, \mathcal{G})$; then φ is the unique homomorphism that extends u.   □

7.4.4 Homology

The boundary map $\partial \colon C_n(S/B) \longrightarrow C_{n-1}(S/B)$ is inherited from $C_n(S)$.

Proposition 7.4.4 ([34], Prop. 2.4) *If $n = 2, 3, 4$, then there exists a unique homomorphism*

$$\partial\colon C_n(S/B) \longrightarrow C_{n-1}(S/B)$$

such that $\partial\langle a_1, a_2, \ldots, a_n\rangle = 0$ *if* $a_1 a_2 \cdots a_n \notin B$, *otherwise*

$$\partial\langle a, b\rangle = \langle b\rangle - \langle ab\rangle + \langle a\rangle \qquad\qquad \textit{if } n = 2,$$

$$\partial\langle a, b, c\rangle = \langle b, c\rangle - \langle ab, c\rangle + \langle a, bc\rangle - \langle a, b\rangle \qquad \textit{if } n = 3,$$

$$\partial\langle a, b, c, d\rangle = \langle b, c, d\rangle - \langle ab, c, d\rangle$$
$$+ \langle a, bc, d\rangle - \langle a, b, cd\rangle + \langle a, b, c\rangle \qquad \textit{if } n = 4,$$

for all $a, b, c, d \in S$. *Moreover,* $\partial \circ \partial = 0$.   □

As in the proof of Proposition 7.2.5 the elements $\partial\langle a_1, \ldots, a_n\rangle$ of $C_{n-1}(S/B)$, as defined above, have property (Cn), so that there exists a unique homomorphism ∂ with values as in the statement. On the other hand, some of the terms in the expansion of $\partial\langle a_1, a_2, \ldots, a_n\rangle$ may disappear: for instance, if $ab \in B$ but $a, b \notin B$, then $\partial\langle a, b\rangle = -\langle ab\rangle$. Several cases must therefore be considered when proving that $\partial \circ \partial = 0$; see [34] for the details.

We now have a complex

$$C_*(S/B)\colon 0 \longleftarrow C_1(S/B) \stackrel{\partial}{\longleftarrow} \cdots \stackrel{\partial}{\longleftarrow} C_4(S/B)$$

of free abelian groups. The homology groups of this complex are the *homology groups of S relative to B*. In detail, let $n = 1, 2, 3$. A *symmetric n-cycle of S relative to B* is an element of

$$Z_n(S/B) = \operatorname{Ker}\partial \subseteq C_n(S/B);$$

a *symmetric n-boundary of S relative to B* is an element of

$$B_n(S/B) = \operatorname{Im}\partial \subseteq Z_n(S/B);$$

and the *n-th homology group of S relative to B* is

$$H_n(S/B) = Z_n(S/B) / B_n(S/B).$$

For example, if S is a monoid and $S \backslash B$ is an ideal of S, then $H_1(S/B)$ is the universal abelian group of the commutative partial monoid B ([34], Prop. 2.5).

If $B = S$, then $C_*(S/B)$ becomes

$$C_*(S)\colon 0 \longleftarrow C_1(S) \stackrel{\partial}{\longleftarrow} \cdots \stackrel{\partial}{\longleftarrow} C_4(S);$$

'*relative to B*' is dropped from the definitions of n-chains, n-cycles, and n-boundaries; $C_n(S/B)$, $Z_n(S/B)$, and $B_n(S/B)$ become $C_n(S)$, $Z_n(S)$, and $B_n(S)$; and $H_n(S/B)$ becomes

$$H_n(S) \;=\; Z_n(S) \,/\, B_n(S),$$

the *n-th homology group of* S (and that's no B S). For example, $H_1(S)$ is the universal abelian group of S.

The homology groups of S are not of much use if S has a zero element:

Proposition 7.4.5 ([34], Prop. 2.6) *If* S *is cursed with a zero element, then* $H_n(S) = 0$ *for* $n = 1, 2, 3$. $\qquad\square$

Proposition 7.4.5 suggests that $H_n(S/\, S\backslash\{0\})$ is a better choice than $H_n(S)$ if S has a zero element; for instance, $H_1(S/\, S\backslash\{0\})$ is the universal abelian group of $S\backslash\{0\}$.

7.4.5   Cohomology

The isomorphism U in Proposition 7.4.3 has the following property:

Lemma 7.4.2 ([34], Lemma 2.7) *If* $n = 1, 2, 3, 4$ *and* $\mathcal{G} = \mathcal{F}(G/B)$ *is semiconstant, then the following square commutes:*

$$
\begin{array}{ccc}
\mathrm{Hom}\,(C_{n-1}(S/B), G) & \xrightarrow{\;\partial^*\;} & \mathrm{Hom}_{\mathcal{T}}\,(C_n(S/B), G) \\[2pt]
{\scriptstyle U}\downarrow & & \downarrow{\scriptstyle U} \\[2pt]
C_S^{n-1}(S, \mathcal{G}) & \xrightarrow{\;\;\delta\;\;} & C_S^{n}(S, \mathcal{G})
\end{array}
$$

where U *is the isomorphism in Proposition 7.4.3 and* $\partial^*\tau = \tau \circ \partial$. $\qquad\square$

Lemma 7.4.2 implies:

Theorem 7.4.1 ([34], Theorem 2.8) *If* $n = 1, 2, 3$ *and* $\mathcal{G} = \mathcal{F}(G/B)$ *is semiconstant, then there is an isomorphism*

$$H_S^n(S, \mathcal{G}) \;\cong\; H^n(C_*(S/B), G)$$

which is natural in G. In particular, if $\mathcal{G} = \mathcal{F}(G/S)$ *is constant, then there is an isomorphism*

$$H_S^n(S, \mathcal{G}) \;\cong\; H^n(C_*(S), G)$$

which is natural in G. $\qquad\square$

Since $C_n(S/B)$ is a free abelian group when $n = 0, 1, 2, 3, 4$, and subgroups of free abelian groups are also free, Theorem 7.4.1 yields a universal coefficients theorem:

Theorem 7.4.2 ([34], Theorem 3.1) *If $n = 1, 2, 3$ and $\mathcal{G} = \mathcal{F}(G/B)$ is semiconstant, then there is an isomorphism*

$$H_S^n(S, \mathcal{G}) \cong \operatorname{Ext}(H_{n-1}(S/B), G) \oplus \operatorname{Hom}(H_n(S/B), G)$$

which is natural in G. In particular, if $\mathcal{G} = \mathcal{F}(G/S)$ is constant, then there is an isomorphism

$$H_S^n(S, \mathcal{G}) \cong \operatorname{Ext}(H_{n-1}(S), G) \oplus \operatorname{Hom}(H_n(S), G)$$

which is natural in G.

Proof This follows from Theorem 7.4.1 and more general Universal Coefficients Theorems, such as Theorem 3.6.5 of [53]. □

Corollaries of Theorem 7.4.2 listed in [34] include:

Proposition 7.4.6 *If S has a zero element and $\mathcal{G}$ is constant, then $H_S^n(S, \mathcal{G}) = 0$ for $n = 1, 2, 3$.*

Proof This follows from Proposition 7.4.5. □

Proposition 7.4.7 ([34], Cor. 3.4) *If S is free, then $H_2(S) = H_3(S) = 0$.* □

Chapter 8
Inheritance

Inheritance is the process by which symmetry properties of cochains induce symmetry properties of their coboundaries. Following [35], this chapter gives a formal definition of symmetry properties, determines all symmetry properties of the coboundaries of symmetric n-cochains when $n \leq 4$, and concludes with conjectures about the appropriate symmetry conditions for dimensions 5 and 6.

8.1 The Universal Coboundary

This section defines symmetry properties and a wondrous universal coboundary, then determines some of the inheritance of (S4).

8.1.1 Symmetry Properties

Let X be a symmetric set of order n, and let $f : X \longrightarrow G$ be a mapping of X into an abelian group G. A *permuted value* of f at $(x_1, x_2, \ldots, x_n) \in X$ is a value $f(x_{\sigma 1}, x_{\sigma 2}, \ldots, x_{\sigma n})$ of f for some permutation σ of $1, 2, \ldots, n$. A *symmetry property* of f at $(x_1, x_2, \ldots, x_n) \in X$ is an equality with integer coefficients p_σ

$$\sum_{\sigma \in S_n} p_\sigma \, f(x_{\sigma 1}, x_{\sigma 2}, \ldots, x_{\sigma n}) = 0$$

between the permuted values $f(x_{\sigma 1}, x_{\sigma 2}, \ldots, x_{\sigma n})$ of f at $(x_1, x_2, \ldots, x_n)$.

For example, the conditions (Sn) in previous chapters are symmetry properties of symmetric cochains at $a, b, \ldots$, when cochains are viewed as collections of mappings of the symmetric sets X_s.

© The Author(s), under exclusive license to Springer Nature Switzerland AG 2022   129
P. A. Grillet, *The Cohomology of Commutative Semigroups*, Lecture Notes
in Mathematics 2307, https://doi.org/10.1007/978-3-031-08212-2_8

To allow multiple mappings to share the same symmetry property the author devised a fine pedantic definition:

A *symmetry property* of order n is a mapping $P\colon \sigma \longmapsto p_\sigma$ of the symmetric group S_n into $\mathbb{Z}$. If $f\colon X \longrightarrow G$ is a mapping of a symmetric set X of order n into an abelian group G, then f *has property* P at $(x_1, x_2, \ldots, x_n) \in X$, and P is a *symmetry property* of f *at* $(x_1, x_2, \ldots, x_n) \in X$, if and only if $\sum_{\sigma \in S_n} p_\sigma f(x_{\sigma 1}, x_{\sigma 2}, \ldots, x_{\sigma n}) = 0$.

With this priceless definition we see that symmetry properties of order n constitute a free $\mathbb{Z}$-module $\mathbb{P} \cong \mathbb{Z}^{n!}$. This comes in handy when deciding whether a symmetry property follows from others.

To each symmetry property P corresponds a *symmetry condition*, which is satisfied by $f\colon X \longrightarrow G$ if and only if f has property P at *every* $(x_1, x_2, \ldots, x_n) \in X$ (where $f\colon X \longrightarrow G$ is a mapping of a symmetric set X into an abelian group G).

For example, all parts of the conditions (Sn) in previous chapters are symmetry conditions for n-cochains, when cochains are viewed as collections of mappings of the symmetric sets X_s.

A symmetry property P of order n is *inherited from* a set (S) of symmetry conditions of order $n - 1$ if and only if it is a property of all coboundaries of $(n - 1)$-cochains that satisfy (S).

8.1.2   The Universal Coboundary

We shall determine *all* the symmetry properties that are inherited from (Sn), where $n \leq 4$.

What makes this is at all possible are miraculous universal coboundaries. As in [35] we construct the universal 5-coboundary, the other universal coboundaries being similar. Let $\mathbb{F}$ be the free commutative semigroup on a set $\{X_1, \ldots, X_5\}$ of 5 distinct indeterminates. Let i denote the canonical ι mapping

$$\mathrm{i}(A, B, C, D) = \langle A, B, C, D \rangle \in C_4(\mathbb{F}).$$

The symmetric mapping i is also a symmetric 4-cochain on $\mathbb{F}$ with values in $C_4(\mathbb{F})$ (more precisely, in the corresponding constant abelian group valued functor on $\mathbb{F}$.) The *universal 5-coboundary* is $\eth = \delta\,\mathrm{i}$.

By definition,

$$\eth(A, B, C, D, E) = \langle B, C, D, E \rangle - \langle AB, C, D, E \rangle + \langle A, BC, D, E \rangle$$
$$- \langle A, B, CD, E \rangle + \langle A, B, C, DE \rangle - \langle A, B, C, D \rangle \in C_4(\mathbb{F}),$$

for all $A, B, C, D, E \in \mathbb{F}$. Properties (C4b) and (C4a) then yield the longer expansion which is used hereafter:

$$\eth(A, B, C, D, E) = \langle B, C, D, E \rangle - \langle AB, C, D, E \rangle \qquad (8.1)$$
$$+ \langle BC, A, D, E \rangle - \langle BC, D, A, E \rangle + \langle BC, D, E, A \rangle$$
$$+ \langle CD, E, B, A \rangle - \langle CD, B, E, A \rangle + \langle CD, B, A, E \rangle$$
$$- \langle DE, C, B, A \rangle - \langle A, B, C, D \rangle.$$

8.1.3 The Group $\mathbb{D}$

To prove the main property of $\eth$ we construct a group $\mathbb{D}$ that contains all the permuted values of $\eth$ at $X_1, \ldots, X_5$. Let

$$T' = \{X_1, \ldots, X_5\},$$
$$T'' = \{X_i X_j \in \mathbb{F} \mid 1 \le i < j \le 5\}, \quad \text{and}$$
$$T = T' \cup T'' \subseteq \mathbb{F}.$$

Totally order $\mathbb{F}$ so that

$$AB < X_1 < X_2 < \cdots < X_5$$

for all $AB \in T''$. Let Y be the standard basis of $\mathbb{F}^4$, which consists of all $(A, B, C, D) \in \mathbb{F}^4$ such that either $A < B, C, D$, or $A \le B, C$ and $B < D$, or both.

Let $Q' \subseteq Y$ be the set of all $(A, B, C, D) \in Y$ such that A, B, C, D are distinct elements of T' (so that either $A = X_1$, or $A = X_2$ and B, C, D is a permutation of X_3, X_4, X_5).

Let Q'' be the set of all $(AB, C, D, E) \in \mathbb{F}^4$ such that $AB \in T''$ and A, B, C, D, E are distinct elements of T' (constitute a permutation of $X_1, \ldots, X_5$).

Let $Q = Q' \cup Q''$, and let $\mathbb{D}$ be the subgroup of $C_4(\mathbb{F})$ generated by $i(Q)$.

Lemma 8.1.1 $\langle A, B, C, D \rangle \in \mathbb{D}$ whenever A, B, C, D are distinct elements of T'.

Proof The set

$$X = \{(A, B, C, D) \in \mathbb{F}^4 \mid A, B, C, D \text{ are distinct elements of } T'\}$$

is a symmetric set. By Proposition 7.2.3, its standard basis is $Y \cap X$. But $Y \cap X = Q'$. Hence $i(Q')$ is a basis of $C(X)$ and $C(X) \subseteq \mathbb{D}$. In particular, $\langle A, B, C, D \rangle \in \mathbb{D}$ for every $(A, B, C, D) \in X$. (This can also be proved directly, using (C4a) and (C4b).) $\qquad\square$

Lemma 8.1.2 $\mathbb{D}$ *is a free abelian group;* $Q \subseteq Y$; $i(Q)$ *is a basis of* $\mathbb{D}$; *and* $\mathbb{D}$ *contains every permuted value of* $\eth$ *at* $X_1, \ldots, X_5$.

Proof By Proposition 7.2.4, $C_4(\mathbb{F})$ is a free abelian group, and $i(Y)$ is a basis of $C_4(\mathbb{F})$. The total order on $\mathbb{F}$ ensures that Y contains both Q' and Q''. Therefore $\mathbb{D}$ is a free abelian group and $i(Q) \subseteq i(Y)$ is a basis of $\mathbb{D}$.

Let $\eth(X_{\sigma 1}, \ldots, X_{\sigma 5})$ be a permuted value of $\eth$. The expansion (8.1) of $\eth(X_{\sigma 1}, \ldots, X_{\sigma 5})$ contains two kinds of terms. Terms $\langle AB, C, D, E \rangle$, where $A, \ldots, E$ is a permutation of $X_1, \ldots, X_5$, all belong to $i(Q'') \subseteq \mathbb{D}$. By Lemma 8.1.1, $\mathbb{D}$ also contains the remaining two terms, $\pm \langle A, B, C, D \rangle$, where A, B, C, D are distinct elements of T'.   □

The main property of $\eth$ is that it maps onto every 5-coboundary:

Lemma 8.1.3 *For every* $u \in C_S^4(S, \mathcal{G})$ *and* $a_1, a_2, \ldots, a_5 \in S$ *there exists a homomorphism* $\varphi : \mathbb{D} \longrightarrow G_s$, *where* $s = a_1 a_2 \cdots a_5$, *such that*

$$\varphi\big(\eth(X_{\sigma 1}, X_{\sigma 2}, \ldots, X_{\sigma 5})\big) = (\delta u)(a_{\sigma 1}, a_{\sigma 2}, \ldots, a_{\sigma 5})$$

for every permutation σ *of* $1, 2, \ldots, 5$.

Proof Since $X_1, X_2, \ldots, X_5$ are distinct there exists a mapping $f : T' \longrightarrow S$ such that $f(X_i) = a_i$ for all i.

In the free commutative semigroup $\mathbb{F}$, $X_i X_j = X_k X_\ell$ implies either $X_i = X_k$ and $X_j = X_\ell$, or $X_j = X_k$ and $X_i = X_\ell$; in either case, $a_i a_j = a_k a_\ell$. Therefore there is a mapping $g : Q'' \longrightarrow S$ such that $g(X_i X_j) = a_i a_j$ for all $1 \le i < j \le 5$.

By Theorem 7.2.1 in Chap. 7, the mapping i is injective on the standard basis Y of $\mathbb{F}^4$; hence $\langle A_1, \ldots, A_4 \rangle = \langle B_1, \ldots, B_4 \rangle$ implies $A_i = B_i$ for all i, when $(A_1, \ldots, A_4), (B_1, \ldots, B_4) \in Y$. In particular, $i(Q')$ and $i(Q'')$ are disjoint. Hence a mapping $h : i(Q) \longrightarrow G_s$, where $s = a_1 a_2 \cdots a_5$, is well defined as follows: if $(A, B, C, D) \in Q'$, then A, B, C, D are all but one (say, X_i) of $X_1, \ldots, X_5$, arranged in a different order, and

$$h\langle A, B, C, D \rangle = u\big((f(A), f(B), f(C), f(D))^{f(X_i)}\big) \in G_s;$$

if $(AB, C, D, E) \in Q''$, then A, B, C, D, E is a permutation of X_1, X_2, X_3, X_4, X_5; $g(AB) = f(A) f(B)$; and

$$h\langle AB, C, D, E \rangle = u\big(f(A) f(B), f(C), f(D), f(E)\big) \in G_s.$$

Since $i(Q)$ is a basis of $\mathbb{D}$, h extends to a homomorphism $\varphi : \mathbb{D} \longrightarrow G_s$.

Now let σ be a permutation of $1, 2, \ldots 5$. Let

$$A, B, C, D, E = X_{\sigma 1}, X_{\sigma 2}, \ldots, X_{\sigma 5} \text{ and } a, b, c, d, e = a_{\sigma 1}, a_{\sigma 2}, \ldots, a_{\sigma 5},$$

so that $f(A) = a$, $f(B) = b$, $f(C) = c$, $f(D) = d$, $f(E) = e$, and $s = abcde$. We have

$$\varphi \langle B, C, D, E \rangle = h \langle B, C, D, E \rangle = u(b, c, d, e)^a \in G_s,$$
$$\varphi \langle AB, C, D, E \rangle = h \langle AB, C, D, E \rangle = u(ab, c, d, e),$$
$$\varphi \langle BC, A, D, E \rangle = h \langle BC, A, D, E \rangle = u(bc, a, d, e),$$
$$\ldots$$
$$\varphi \langle DE, C, B, A \rangle = h \langle DE, C, B, A \rangle = u(de, c, b, a), \quad \text{and}$$
$$\varphi \langle A, B, C, D \rangle = h \langle A, B, C, D \rangle = u(a, b, c, d)^e \in G_s.$$

Hence

$$\begin{aligned}
\varphi \big(\eth(A, B, C, D, E) \big) &= \varphi \langle B, C, D, E \rangle - \varphi \langle AB, C, D, E \rangle \\
&\quad + \varphi \langle BC, A, D, E \rangle - \varphi \langle BC, D, A, E \rangle + \varphi \langle BC, D, E, A \rangle \\
&\quad + \varphi \langle CD, E, B, A \rangle - \varphi \langle CD, B, E, A \rangle + \varphi \langle CD, B, A, E \rangle \\
&\quad - \varphi \langle DE, C, B, A \rangle - \varphi \langle A, B, C, D \rangle \\
&= u(b, c, d, e)^a - u(ab, c, d, e) \\
&\quad + u(bc, a, d, e) - u(bc, d, a, e) + u(bc, d, e, a) \\
&\quad + u(cd, e, b, a) - u(cd, b, e, a) + u(cd, b, a, e) \\
&\quad - u(de, c, b, a) - u(a, b, c, d)^e \\
&= (\delta u)(a, b, c, d, e),
\end{aligned}$$

by (S4d) and (S4b). Thus

$$\varphi \big(\eth(X_{\sigma 1}, X_{\sigma 2}, \ldots, X_{\sigma 5}) \big) = (\delta u)(a_{\sigma 1}, a_{\sigma 2}, \ldots, a_{\sigma 5}).$$

$$\square$$

Theorem 8.1.1 *A symmetry property that does not require an equality between its variables is inherited from* (S4) *if and only if it is a property of $\eth$ at $X_1, X_2, \ldots, X_5$.*

Proof If a symmetry property P is inherited from (S4), then P holds at every $a_1, \ldots, a_5$ for the coboundary of every symmetric 4-cochain on any commutative semigroup (without requiring any two of $a_1, \ldots, a_5$ to be equal), and in particular holds for $\delta i = \eth$ at $X_1, \ldots, X_5$. (On the other hand, a symmetry property that requires an equality between its variables cannot apply to $\eth$ at $X_1, \ldots, X_5$, since $X_1, \ldots, X_5$ are distinct.)

Conversely, let $P : \sigma \longmapsto p_\sigma$ hold for $\mathfrak{d}$ at $X_1, \ldots, X_5$, so that

$$\sum_{\sigma \in S_5} p_\sigma \, \mathfrak{d} \, (X_{\sigma 1}, \, X_{\sigma 2}, \ldots, X_{\sigma 5}) = 0 \, .$$

Let u be a 4-cochain on a commutative semigroup S with values in some abelian group valued functor $\mathcal{G} = (G, \gamma)$. Let $a_1, a_2, \ldots, a_5 \in S$ and let $s = a_1 a_2 \cdots a_5$. By Lemma 8.1.3 there is a homomorphism $\varphi : \mathbb{D} \longrightarrow G_s$ such that

$$\varphi \left(\mathfrak{d} \, (X_{\sigma 1}, \, X_{\sigma 2}, \ldots, X_{\sigma 5}) \right) = (\delta u)(a_{\sigma 1}, \, a_{\sigma 2}, \ldots, a_{\sigma 5})$$

for every permutation σ of $1, 2, \ldots, 5$. Then

$$\sum_{\sigma \in S_5} p_\sigma \, (\delta u)(a_{\sigma 1}, \ldots, a_{\sigma 5}) = \varphi \left(\sum_{\sigma \in S_5} p_\sigma \, \mathfrak{d} \, (X_{\sigma 1}, \ldots, X_{\sigma 5}) \right) = 0$$

and P holds for δu at $a_1, a_2, \ldots, a_5$. $\qquad\qquad\qquad\qquad\qquad\qquad \square$

8.2   One Equality Between Variables

Theorem 8.1.1 does not cover symmetry properties that require one or more equalities between its variables, like the symmetry condition (S4c) which requires two equalities between its four variables. If some of $a_1, a_2, \ldots, a_n$ are equal, then there are fewer permutations of $a_1, a_2, \ldots, a_n$, fewer permuted values $f(a_{\sigma 1}, a_{\sigma 2}, \ldots, a_{\sigma n})$, and different relationships between permuted values.

In this section, as in [35], we consider inheritances of (S4) that require exactly one equality between variables, the other cases being similar. In this case a simple tweak of $\mathfrak{d}$ yields a result similar to Theorem 8.1.1.

Let the required equality between a_1, a_2, a_3, a_4 be $a_p = a_q$, where $1 \leq p < q \leq 4$. Using X_1, X_2, X_3, X_4, make a copy of this equality: define $Y_1, Y_2, \ldots, Y_5$ so that $\{Y_1, Y_2, \ldots, Y_5\} = \{X_1, X_2, X_3, X_4\}$ and $Y_p = Y_q = X_1$. Since $X_1, X_2, \ldots, X_5$ are distinct, $Y_p = Y_q$ is the only equality between $Y_1, Y_2, \ldots, Y_5$. For example, if the required equality is $a_2 = a_4$, then X_2, X_1, X_3, X_1, X_4 serves as $Y_1, Y_2, \ldots, Y_5$.

Using a different version of the sets T', T'', Q', and Q'', construct a different version of the free abelian group $\mathbb{D}$ that will contain all permuted values of $\mathfrak{d}$ at $(Y_1, Y_2, \ldots, Y_5)$. Let

$$T' = \{Y_1, \ldots, Y_5\} = \{X_1, X_2, X_3, X_4\},$$

$$T'' = \{Y_i Y_j \in C_4(\mathbb{F}) \mid 1 \leq i < j \leq 5\}, \quad \text{and}$$

$$T = T' \cup T''.$$

Totally order $\mathbb{F}$ as before, so that

$$Y_i Y_j < X_1 < \cdots < X_4$$

for all $1 \le i < j \le 5$. Then $\mathbb{F}^4$ has the same standard basis Y, which consists of all $(A, B, C, D) \in \mathbb{F}^4$ such that either $A < B, C, D$, or $A \le B, C$ and $B < D$, or both.

Let $Q' \subseteq Y$ be the set of all $(A, B, C, D) \in Y$ such that A, B, C, D is a partial permutation of $Y_1, \ldots, Y_5$ (so that $A, B, C, D \in T'$ are distinct except perhaps for one equality).

Let Q'' be the set of all $(AB, C, D, E) \in \mathbb{F}^4$ such that $AB \in T''$ and A, B, C, D, E is a permutation of $Y_1, Y_2, \ldots, Y_5$.

Let $Q = Q' \cup Q''$ and let $\mathbb{D}$ be the subgroup of $C_4(\mathbb{F})$ generated by $i(Q)$.

The new group $\mathbb{D}$ and his brother in Sect. 8.1 have similar properties.

First, $\mathbb{D}$ is a free abelian group; $Y \subseteq Q$; $i(Q)$ is a basis of $\mathbb{D}$; and $\mathbb{D}$ contains every permuted value of $\eth(Y_1, \ldots, Y_5)$. This is proved like Lemmas 8.1.1 and 8.1.2.

Next, for every $u \in C_S^4(S, \mathcal{G})$ and every $a_1, a_2, \ldots, a_5 \in S$ such that $a_p = a_q$, there exists a homomorphism $\varphi : \mathbb{D} \longrightarrow G_s$, where $s = a_1 a_2 \cdots a_5$, such that

$$\varphi\left(\eth(Y_{\sigma 1}, \ldots, Y_{\sigma 5})\right) = (\delta u)(a_{\sigma 1}, \ldots, a_{\sigma 5})$$

for every permutation σ of $1, 2, \ldots, 5$. This is proved like Lemma 8.1.3. Hence

Theorem 8.2.1 *Let* $1 \le p < q \le 5$. *A symmetry property that requires one equality* $a_p = a_q$ *between its variables* $a_1, a_2, \ldots, a_5$, *and does not require any other equality between its variables, is inherited from* (S4) *if and only if it is a property of* $\eth$ *at* $Y_1, Y_2, \ldots, Y_5$.

This is proved like Theorem 8.1.1.

Now $Y_1, Y_2, \ldots, Y_5$ is a permutation of X_1, X_1, X_2, X_3, X_4. Hence the permuted values of $\eth$ at $Y_1, \ldots, Y_5$ are the same as its permuted values at X_1, X_1, X_2, X_3, X_4 (though in a different order); the same relationships exist between these permuted values; and the set of all symmetry properties of $\eth$ at $Y_1, \ldots, Y_5$ is the same as the set of all its symmetry properties at X_1, X_1, X_2, X_3, X_4. Therefore

Theorem 8.2.2 *A symmetry property that requires exactly one equality between its variables is inherited from* (S4) *if and only if it is a property of* $\eth$ *at* X_1, X_1, X_2, X_3, X_4.

8.3 Results

This section determines all symmetry properties that are inherited from (S2), (S3), or (S4).

8.3.1   Method

Based on Theorems 8.1.1 and 8.2.2, a symmetry property of order $n+1$ that requires certain equalities between its variables (perhaps none) is inherited from (Sn) if and only if it is a symmetry property of $\mathfrak{d}$ at some suitable sequence (X) made from $X_1, X_2, \ldots, X_{n+1}$. The permuted values of $\mathfrak{d}$ at (X) all belong to some free $\mathbb{Z}$-module $\mathbb{D}$ with a basis $\mathfrak{i}(Q)$. Linear relationships between permuted values are then found from their coordinate matrix in the basis $\mathfrak{i}(Q)$.

Let $\mathbb{V}$ be the subgroup of $\mathbb{D}$ generated by the permuted values. It turns our that some of the permuted values, the *basic permuted value s*, constitute a basis of $\mathbb{V}$. Such a basis is revealed by Gauss-Jordan reduction of the coordinate matrix. Writing the remaining permuted values as linear combinations of the basic permuted values provides *basic relationships*, and *basic symmetry properties*. Every linear relationship between permuted values is a consequence of these basic relationships; hence every symmetry property of $\mathfrak{d}$ at the sequence (X) is a consequence of the basic symmetry properties.

Row reduction can also reveal if a symmetry property follows from other symmetry properties, or from symmetry conditions. In the free $\mathbb{Z}$-module $\mathbb{P}$ of symmetry properties of order n, if a symmetry property P is a linear combination of other symmetry properties $P_1, \ldots, P_k$, then P is a consequence (as a property) of $P_1, \ldots, P_k$. A set (S) of symmetry condition of order n consists of symmetry properties (one for each symmetry condition and each permutation of the n variables); a symmetry property which is a linear combination of these symmetry properties is a consequence of (S). Linear combinations of symmetry properties are detected by Gauss-Jordan reduction of their coordinate matrix in a basis of $\mathbb{P}$. Thus it can be established that a symmetry property is a consequence of (S).

8.3.2   Order 5

We apply these methods to the inheritance of (S4).

We begin with symmetry properties that require no equalities between variables, which, by Theorem 8.1.1, are the symmetry properties of $\mathfrak{d}$ at $X_1, X_2, \ldots, X_5$. To ease displays and outputs we arranged the coordinates of the permuted values of $\mathfrak{d}(X_1, X_2, \ldots, X_5)$ in rows rather than columns. The resulting matrix has 120 rows, one for each permuted value, and 90 columns, one for each element of Q. Computer row reduction of the coordinate matrix found its rank to be 24 and yielded 96 basic symmetry properties, including

$$\mathfrak{d}(X_5, X_4, X_3, X_2, X_1) = \mathfrak{d}(X_1, X_2, X_3, X_4, X_5) \text{ and}$$

$$\mathfrak{d}(X_2, X_1, X_3, X_4, X_5) = \mathfrak{d}(X_1, X_2, X_3, X_4, X_5) - \mathfrak{d}(X_1, X_3, X_2, X_4, X_5)$$
$$+ \mathfrak{d}(X_1, X_3, X_4, X_2, X_5) - \mathfrak{d}(X_1, X_3, X_4, X_5, X_2).$$

Therefore every 5-coboundary u satisfies the corresponding two symmetry conditions, which form a set (S5):

$$u(e, d, c, b, a) = u(a, b, c, d, e) \text{ and} \tag{S5a}$$

$$u(b, a, c, d, e) = u(a, b, c, d, e) - u(a, c, b, d, e)$$
$$+ u(a, c, d, b, e) - u(a, c, d, e, b). \tag{S5b}$$

Another row reduction established that every basic symmetry property that requires no equality is a consequence of (S5).

Between the five variables a, b, c, d, e of 5-coboundaries a symmetry property can require either no equality, or one equality (e.g. $a = b$), or two equalities (e.g. $a = b$ and $c = d$), or three equalities (e.g. $a = b = c$), or four equalities (e.g. $a = b = c, d = e$), or six equalities (e.g. $a = b = c = d$), or ten equalities ($a = b = c = d = e$). These cases must be considered separately.

By Theorem 8.2.2 the symmetry properties that require exactly one equality are the symmetry properties of $\eth$ at X_1, X_1, X_2, X_3, X_4. The coordinate matrix now has 60 rows, 48 columns, rank 12, and provides 48 basic symmetry properties, all of which are consequences of (S5).

The remaining five cases are treated by the same method using similar results, to obtain:

Theorem 8.3.1 *A symmetry property of order 5 is inherited from* (S4) *if and only if it is a consequence of* (S5).

8.3.3 Other Orders

The same method yields:

Theorem 8.3.2 *A symmetry property of order 3 is inherited from* (S2) *if and only if it is a consequence of* (S3).

Theorem 8.3.3 *A symmetry property of order 4 is inherited from* (S3) *if and only if it is a consequence of* (S4).

In view of this it seems reasonable to conjecture that (S5) is an appropriate set of symmetry conditions for 5-cochains, meaning that it leads to $H_S^5(S, \mathcal{G}) \cong H_B^5(S, \mathcal{G})$. But this is easier stated than proved.

Moreover, (S5) lends itself to standard bases for symmetric sets of order 5. Hence the process can be continued, resulting in:

Theorem 8.3.4 *A symmetry property of order 6 is inherited from* (S5) *if and only if it is a consequence of* (S6), *which consists of four conditions:*

$$u\,(a,b,c,d,e,f)\;=\;u\,(f,e,d,c,b,a)\,, \tag{S6a}$$

$$u\,(a,b,c,d,e,f)\;=\;u\,(b,a,c,d,e,f)\;-\;u\,(b,c,a,d,e,f) \tag{S6b}$$
$$+\;u\,(b,c,d,a,e,f)\;-\;u\,(b,c,d,e,a,f)$$
$$+\;u\,(b,c,d,e,f,a)\,,$$

$$u\,(a,b,c,d,e,f)\;=\;-\;u\,(c,b,a,d,e,f)\;+\;u\,(c,b,d,a,e,f) \tag{S6c}$$
$$-\;u\,(c,b,d,e,a,f)\;+\;u\,(c,b,d,e,f,a)$$
$$-\;u\,(c,d,b,a,e,f)\;+\;u\,(c,d,b,e,a,f)$$
$$-\;u\,(c,d,b,e,f,a)\;-\;u\,(c,d,e,b,a,f)$$
$$+\;u\,(c,d,e,b,f,a)\;-\;u\,(c,d,e,f,b,a)\,,\;\;and$$

$$u\,(a,b,a,b,a,b)\;=\;0\,. \tag{S6d}$$

It seems reasonable to conjecture that (S6) is an appropriate definition of symmetric 6-cochains.

Due to extensive use of computers, Theorems 8.3.1 through 8.3.4 are to a large extent experimental. They have not been independently confirmed.

The author is not afraid of hard work, but there are limits; so he stopped after Theorem 8.3.4.

Chapter 9
Appendixes

9.1 Extensions

This appendix reviews group extensions and provides an overview of three extension theories that generalize group extensions to not necessarily commutative semigroups: Rédei extensions [47], group coextensions [18], and Leech coextensions [39].

9.1.1 Group Extensions

A *group extension* $E = (E, \pi)$ of a group G by a group Q is a group E with a surjective homomorphism $\pi : E \longrightarrow Q$ with kernel G; equivalently, up to isomorphisms, such that G is a normal subgroup of E and $E/G \cong Q$.

Two extensions (E, π) and (E', π') of G by Q are *equivalent* if and only if there exists an isomorphism $\theta : E \longrightarrow E'$ (an *equivalence* of extensions) which is the identity on G ($\theta g = g$ for all $g \in G$) and preserves projection to Q ($\pi' \circ \theta = \pi$).

Schreier's method [52] for constructing any extension $E = (E, \pi)$ of G by Q begins with a choice of $p_a \in E_a = \pi^{-1}\{a\}$ for each $a \in Q$. Then every element x of E can be written in the form $x = g p_a$ for some unique $g \in G$ and $a \in Q$. In particular,

$$p_a\, p_b = s_{a,b}\, p_{ab}$$

for some unique $s_{a,b} \in G$; the family $s = (s_{a,b})_{a,b \in Q}$ is the *factor set* of E. Similarly

$$p_a\, g = {}^a g\, p_a$$

© The Author(s), under exclusive license to Springer Nature Switzerland AG 2022

P. A. Grillet, *The Cohomology of Commutative Semigroups*, Lecture Notes in Mathematics 2307, https://doi.org/10.1007/978-3-031-08212-2_9

for some unique $^a g \in G$, namely $^a g = p_a g \, p_a^{-1}$. Thus the set Q acts on G by automorphisms; but this action is not necessarily a group action. Associativity in E yields:

$$^{a(^b k)} \, ^a s_{b,c} \, s_{a,bc} = s_{a,b} \, ^{ab} k \, s_{ab,c} \tag{9.1}$$

for all $a, b, c \in Q$ and $k \in G_c$. It can be arranged that $p_1 = 1$, and then $^1 g = g$ and s is *normalized* ($s_{1,a} = s_{a,1} = 1$ for all a).

Elements of E then multiply by

$$(g p_a)(h p_b) = g \, ^a h \, s_{a,b} \, p_{ab} \,. \tag{9.2}$$

Conversely, if (9.1) holds, then $G \times Q$, with g identified with $(g, 1)$, projection $(g, a) \mapsto a$ to Q, and multiplication

$$(g, a)(h, b) = (g \, ^a h \, s_{a,b} \,, \; ab) \,, \tag{9.3}$$

is an extension of G by Q; and every extension of G by Q is equivalent to an extension constructed in this fashion [52]. Moreover, two extensions of G by Q with factor sets s and s' are equivalent if and only if there exists $u = (u_a)_{a \in Q}$ such that $u_a \in G_a$ and

$$s'_{a,b} \, u_{ab} = u_a \, ^a u_b \, s_{a,b} \,, \tag{9.4}$$

for all $a, b \in Q$.

Conditions (9.1) and (9.4) become simpler if G is abelian. With G written additively, (9.1) splits into two conditions:

$$^{a(^b k)} = {}^{ab} k \,, \tag{9.5}$$

so that the action of Q on G is a group action, which makes G a Q-module (or a $\mathbb{Z}[Q]$-module); and

$$^a s_{b,c} - s_{ab,c} + s_{a,bc} - s_{a,b} = 0 \,. \tag{9.6}$$

Moreover, two extensions of G by Q with factor sets s and s' are equivalent if and only if

$$s'_{a,b} - s_{a,b} = {}^a u_b - u_{ab} + u_a \tag{9.7}$$

for some $u_a \in G_a$, one for each $a \in Q$. These last two conditions lead directly to the cohomology of the group Q (see e.g. [41]) with coefficients in G, in which a (normalized) n-cochain u assigns $u(a_1, a_2, \ldots, a_n) \in G$ to each $a_1, a_2, \ldots, a_n \in Q$, so that $u(a_1, a_2, \ldots, a_n) = 0$ if $a_i = 1$ for any i, with coboundary%

$$(\partial u)(a_1, a_2, \ldots, a_{n+1}) = {}^{a_1}u(a_2, \ldots, a_{n+1}) \tag{9.8}$$
$$+ \sum_{1 \leq i \leq n} (-1)^i u(a_1, \ldots, a_i a_{i+1}, \ldots, a_{n+1})$$
$$+ (-1)^{n+1} u(a_1, a_2, \ldots, a_n).$$

Then (9.6) states that s is a 2-cocycle and (9.7) states that two factor sets are equivalent if and only if they differ by a 2-coboundary. Hence the cohomology group $H^2(Q, G)$ *classifies* group extensions of S by Q: there is a one-to-one correspondence between the elements of $H^2(Q, G)$ and the equivalence classes of group extensions of G by Q [41].

9.1.2 Rédei Extensions

Rédei extensions [47] are a direct generalization of group extensions. A monoid E is a *Rédei extension* of a monoid M by a monoid Q if and only if M is a submonoid of E, there exists a homomorphism π of E onto Q, and for each $a \in Q$ there exists $p_a \in E_a = \pi^{-1}\{a\}$ such that $p_1 = 1 \in M$ and every element x of E can be written uniquely in the form $x = m p_a$ with $m \in M$ and $a \in Q$ (in the form $x = am$ in the original definition [47]). In particular, $M = \pi^{-1}\{1\}$.

Schreier's construction can be applied directly to Rédei extensions. If E is a Rédei extension of M by Q, then, for the given choice of p, E has a normalized factor set s defined by $p_a p_b = s_{a,b} p_{ab}$, and an action of the set Q on M defined by $p_a g = {}^a g \, p_a$, that together satisfy (9.1); so that (9.2) yields products in E and E is equivalent to $M \times Q$ with the multiplication in (9.3).

If $M = G$ is an abelian group, then, as before, (9.1) splits into (9.5) and (9.6), so that G becomes a Q-module (or a $\mathbb{Z}[Q]$-module). This leads to the *Eilenberg-MacLane cohomology* of the monoid Q, in which an n-cochain u assigns $u(a_1, a_2, \ldots, a_n) \in G$ to each $a_1, a_2, \ldots, a_n \in Q$, with coboundary (9.8). Then factor sets are 2-cocycles, two factor sets are equivalent if and only if they differ by a 2-coboundary, and $H^2(Q, G)$ classifies Rédei extensions of G by Q.

9.1.3 The Leech Categories

Unfortunately, it is a rare monoid that can be constructed as a Rédei extension. To reach every semigroup, the single monoid M must be replaced by several groups, one group for each set E_a, which acts simply and transitively on E_a. These groups are organized into functors on one of the Leech categories [39].

The *Leech category* $\mathcal{R}(S)$ of a semigroup S (denoted by $\mathbb{R}(S)$ in [39]) is the small category whose objects are the elements of S, with one morphism $(a, t): a \longrightarrow at$ for each $a \in S$ and $t \in S^1$. Composition is by $(at, u) \circ (a, t) = (a, tu)$. The identity on $a \in S$ is $(a, 1)$. A group valued functor $\mathcal{G} = (G, \gamma)$ on $\mathcal{R}(S)$ thus assigns a group G_a to each $a \in S$ and a homomorphism $\gamma_{a,t}: G_a \longrightarrow G_{at}$ for each $a \in S$ and $t \in S^1$. It is convenient to denote $\gamma_{a,t}\, g$ by g^t, so that

$$(gh)^t = g^t\, h^t, \quad g^1 = g, \quad \text{and} \quad (g^t)^u = g^{tu}$$

for all g, h, t, u.

Dually, the *Leech category* $\mathcal{L}(S)$ of a semigroup S (denoted by $\mathbb{L}(S)$ in [39]) is the small category whose objects are the elements of S, with one morphism $(t, a): a \longrightarrow ta$ for each $a \in S$ and $t \in S^1$. Composition is by $(t, ua) \circ (u, a) = (tu, a)$. The identity on $a \in S$ is $(1, a)$. A group valued functor $\mathcal{G} = (G, \gamma)$ on $\mathcal{L}(S)$ thus assigns a group G_a to each $a \in S$ and a homomorphism $\gamma_{t,a}: G_a \longrightarrow G_{ta}$ for each $a \in S$ and $t \in S^1$. It is convenient to denote $\gamma_{t,a}\, g$ by $^t g$. Then

$$^t(gh) = {}^t g\, {}^t h, \quad ^1 g = g, \quad \text{and} \quad ^t(^u g) = {}^{tu} g$$

for all g, h, t, u.

If S is commutative, then $\Theta: a \mapsto a$, $(a, t) \mapsto (t, a)$ is an isomorphism of categories of $\mathcal{R}(S)$ onto $\mathcal{L}(S)$. Then we denote $\mathcal{R}(S)$ by $\mathcal{H}(S)$. (No $\mathbb{H}(S)$ category is defined in [39].)

The two-sided *Leech category* $\mathcal{D}(S)$ of a semigroup S (denoted by $\mathbb{D}(S)$ in [39]) is the small category whose objects are the elements of S, with one morphism $(t, a, u): a \longrightarrow tau$ for each $a \in S$ and $t, u \in S^1$. Composition is by $(s, tau, v) \circ (t, a, u) = (st, a, uv)$. The identity on $a \in S$ is $(1, a, 1)$. A group valued functor $\mathcal{G} = (G, \gamma)$ on $\mathcal{D}(S)$ thus assigns a group G_a to each $a \in S$ and a homomorphism $\gamma_{t,a,u}: G_a \longrightarrow G_{tau}$ for each $a \in S$ and $t, u \in S^1$. It is convenient to denote $\gamma_{t,a,u}\, g$ by $^t g^u$. Then

$$^t(gh)^u = {}^t g^u\, {}^t h^u, \quad ^1 g^1 = g, \quad \text{and} \quad ^s(^t g^u)^v = {}^{st} g^{uv}$$

for all g, s, t, u, v.

A group valued functor $\mathcal{G} = (G, \gamma)$ on $\mathcal{D}(S)$ induces group valued functors on $\mathcal{R}(S)$ and $\mathcal{L}(S)$, in which $g^t = {}^1 g^t$ and $^t g = {}^t g^1$ for all g, t, u. Then $^t g^u = (^t g)^u = {}^t(g^u)$.

9.1.4 Cosets

Groups that act simply and transitively on subsets of S are not uncommon, as shown by the Schützenberger groups of $\mathcal{H}$-class es. More generally, a *left coset* of a semigroup S is a nonempty subset C of S with the following properties: if $c, d \in C$, then $d = tc$ for some $t \in S^1$; if $c \in C$, $t \in S^1$, and $tc \in C$, then $tC \subseteq C$; if $c, d \in C$, $t \in S^1$, and $tc = td$, then $c = d$ [18]. The dual properties define a *right coset* of S.

If S is a group and G is a subgroup of S, then every left coset Ga of G is a left coset of S. In any semigroup S, every $\mathcal{H}$-class is a left coset. Moreover, every congruence class that is contained in a single $\mathcal{H}$-class is both a left coset of S and a right coset of S [18].

If C is a left coset of S, every element of

$$C^L = \{ t \in S^1 \mid tC \subseteq C \}$$

induces a transformation

$$\widehat{t} : c \mapsto tc, \quad C \longrightarrow C$$

of C. Then

$$G^L(C) = \{ \widehat{t} \mid t \in C^L \}$$

is a group of transformations of C, the *left Schützenberger group* of C (called the *generalized left Schützenberger group* of C in [18]), which acts simply and transitively on C by: $\widehat{t} \cdot c = tc$. Dually, every right coset C of S has a *right Schützenberger group*, which acts simply and transitively on C on the right by: $c \cdot \widehat{u} = cu$.

Homomorphisms between Schützenberger groups arise when two left cosets C and D satisfy $Cu \subseteq D$ for some $u \in S^1$. Then $C^L \subseteq D^L$ and the inclusion map $C^L \longrightarrow D^L$ induces a homomorphism $\gamma_D^C : G^L(C) \longrightarrow G^L(D)$, unique since $C^L \longrightarrow G^L(C)$ is surjective, such that the following square commutes:

$$
\begin{array}{ccc}
C^L & \longrightarrow & G^L(C) \\
\subseteq \downarrow & & \downarrow \gamma_D^C \\
D^L & \longrightarrow & G^L(D)
\end{array}
\tag{9.9}
$$

In particular, γ_D^C does not depend on u. Since $g \in G^L(C)$ and $\gamma_D^C g \in G^L(D)$ are induced by the same $t \in C^L$, we have

$$(g \cdot c)\, u = (tc)\, u = t\, (cu) = \gamma_D^C g \cdot cu \tag{9.10}$$

for all $c \in C$. Uniqueness in (9.9) implies $\gamma_C^C = 1$ and $\gamma_E^D \circ \gamma_D^C = \gamma_E^C$ if $Cu \subseteq D$ and $Dv \subseteq E$ for some u, v.

A *left coset congruence* is a congruence $\mathcal{C}$ such that every $\mathcal{C}$-class is a left coset. In particular, $\mathcal{C}$ is contained in Green's relation $\mathcal{L}$, but not necessarily contained in $\mathcal{H}$ [18].

If $\mathcal{C}$ is a left coset congruence, then the left Schützenberger groups of the $\mathcal{C}$-classes become values of a functor, the *left Schützenberger functor* $\mathcal{G}^L(\mathcal{C}) = (G^L, \gamma^L)$ of $\mathcal{C}$, which is a group valued functor on $\mathcal{R}(S/\mathcal{C})$. Let $C \in S/\mathcal{C}$, let $U \in (S/\mathcal{C})^1$, and let $u \in U$; in what follows we regard $u \in U$ as meaning $u = 1 \in S^1$ in case $S/\mathcal{C}$ does not have an identity element and $U = 1 \notin S/\mathcal{C}$. Then $Cu \subseteq CU$, the product of C and U in $(S/\mathcal{C})^1$, and there is a homomorphism $\gamma_{C,U}^L = \gamma_{CU}^C : G^L(C) \longrightarrow G^L(CU)$; by (9.10),

$$(g \cdot c) u = (\gamma_{C,U}^L g) \cdot cu$$

for all $g \in G^L(C)$, $c \in C$, and $u \in U$. This left Schützenberger functor is *thin* in that $CU = CV$ implies $\gamma_{C,U}^L = \gamma_{C,V}^L$.

If S is commutative, then so is C^L, and all Schützenberger groups are abelian. Moreover, every left coset congruence is contained in $\mathcal{H}$.

9.1.5 Group Coextensions

General group coextensions encapsulate properties of left coset congruences and their Schützenberger functors [18]. A *group coextension* $E = (E, \pi, \cdot)$ of a semigroup S by a group valued functor $\mathcal{G} = (G, \gamma)$ on $\mathcal{R}(S)$ consists of a semigroup E, a homomorphism π of E onto S, and, for each $a \in S$, a simply transitive action $\cdot$ of G_a on $E_a = \pi^{-1}\{a\}$, such that

$$(g \cdot x) y = g^b \cdot xy$$

whenever $g \in G_a$, $x \in E_a$, $y \in E_b$, and $a, b \in S$.

If $\mathcal{C}$ is a left coset congruence on a semigroup S, then S is a group coextension of $S/\mathcal{C}$ by the left Schützenberger functor of $\mathcal{C}$. A Rédei extension of a group G by a monoid Q is a group coextension of Q by the constant group valued functor on $\mathcal{R}(Q)$ that assigns G to every $a \in Q$ and the identity on G to every $(a, t) \in Q \times Q$. In particular, a group extension of G by Q is a group coextension of Q by G.

Two group coextensions E and E' of S by $\mathcal{G}$ are *equivalent* if and only if there exists an isomorphism (an *equivalence* of group coextension s) $\theta : E \longrightarrow E'$ which preserves projection to S: $\theta(E_a) = E'_a$ and the actions of $\mathcal{G}$: $\theta(g \cdot x) = g \cdot \theta x$, for all $g \in G_a$, $x \in E_a$, and $a \in S$.

Group coextensions are constructed, up to equivalence, by Schreier's method. Choose $p_a \in E_a$ for each $a \in S$. Since G_a acts simply and transitively on E_a there is for each $x \in E_a$ a unique $g \in G_a$ such that $x = g \cdot p_a$. In particular,

$$p_a \, p_b = s_{a,b} \cdot p_{ab}$$

for some unique $s_{a,b} \in G_{ab}$; s is the *factor set* of E. Moreover,

$$p_a \, (g \cdot p_b) = {}^a_* g \cdot (p_a \, p_b) \tag{9.11}$$

for some unique ${}^a_* g \in G_{ab}$ (which also depends on b). Regrettably, $g \mapsto {}^a_* g$ need not be a homomorphism [18]. Products in E are now completely determined by the factor set s and commutation operation $a, g \mapsto {}^a_* g$:

$$(g \cdot p_a)(h \cdot p_b) = (g^b \, {}^a_* h \, s_{a,b}) \cdot p_{ab} \, . \tag{9.12}$$

Associativity in E yields the property

$$({}^a_* h)^c \, s^c_{a,b} \, s_{ab,c} \, {}^{ab}_* k = {}^a_* (h^c \, {}^b_* k \, s_{b,c}) \, s_{a,bc} \tag{9.13}$$

for all a, b, c, h, k.

Conversely, if (9.13) holds, then

$$E[S, \mathcal{G}, s, *] = \{ (g, a) \mid a \in S, \; g \in G_a \},$$

with multiplication

$$(g, a)(h, b) = (g^b \, {}^a_* h \, s_{a,b}, \; ab), \tag{9.14}$$

projection $(g, a) \mapsto a$ to S, and action $g \cdot (h, a) = (gh, a)$ of G_a on E_a for each $a \in S$, is a group coextension of S by $\mathcal{G}$; and every group coextension of S by $\mathcal{G}$ is equivalent to some $E[S, \mathcal{G}, s, *]$.

For later comparisons, products (9.12) and (9.14) can be put into sandwich form: let ${}^a h = s^{-1}_{a,b} \, {}^a_* h \, s_{a,b}$; then

$$(g, a)(h, b) = (g^b \, s_{a,b} \, {}^a h, \; ab) \, . \tag{9.15}$$

Finally, let E be a group coextension of S by a group valued functor $\mathcal{G}$. If E is commutative, then S is commutative and (9.11) yields

$${}^a_* g \cdot p_a \, p_b = p_a \, (g \cdot p_b) = (g \cdot p_b) \, p_a = g^a \cdot p_a \, p_b \, ,$$

so that ${}^a_* g = g^a$ whenever $g \in G_a$; in particular, $g \mapsto {}^a_* g$ is a homomorphism and ${}^a_* g$ depends only on g and a. Then it follows from (9.12) that E is commutative if and only if S is commutative and, for all $a, b \in S$, $g \in G_a$, and $h \in G_b$,

$$g^b \, h^a \, s_{a,b} = h^a \, g^b \, s_{b,a} \, ;$$

then

$$s_{b,a} = s_{a,b} \tag{9.16}$$

for all a, b. Conversely, if S is commutative and all groups G_a are abelian, then (9.16) implies that E is commutative; then we recognize in E a commutative group coextension as defined in Chap. 1.

9.1.6 Congruences Contained in $\mathcal{H}$

These are in fact the most interesting instances of left coset congruences. A congruence $\mathcal{C}$ contained in Green's relation $\mathcal{H}$ is both a left coset congruence and a right coset congruence and has a left Schützenberger functor $\mathcal{G}^L(\mathcal{C}) = (G^L, \gamma^L)$, which is a group valued functor on $\mathcal{R}(S/\mathcal{C})$, and a right Schützenberger functor $\mathcal{G}^R(\mathcal{C}) = (G^R, \gamma^R)$, which is a group valued functor on $\mathcal{L}(S/\mathcal{C})$.

We note three properties. First, the actions of $G^L(C)$ and $G^R(C)$ on each $\mathcal{C}$-class C commute: if $g \in G^L(C)$ is induced by $t \in C^L$ and $h \in G^R(C)$ is induced by $u \in C^R$, then associativity in S yields

$$(g \cdot c) \cdot h = (tc)u = t(cu) = g \cdot (c \cdot h) \tag{9.17}$$

for all $c \in C$.

Second, $G^L(C)$ and $G^R(C)$ are isomorphic: for any $p \in C$, there is for each $g \in G^L(C)$ a unique $g' \in G^R(C)$ such that $g \cdot p = p \cdot g'$; then

$$gh \cdot p = g \cdot (h \cdot p) = g \cdot p \cdot h' = (p \cdot g') \cdot h' = p \cdot g'h',$$

by (9.17), so that $\theta_p : g \mapsto g'$, $G^L(C) \longrightarrow G^R(C)$ is an isomorphism.

Third, the right action of $G^R(C)$ on C is determined by the left action of $G^L(C)$: for all $h \in G^L(C)$ and $x = g \cdot p \in C$,

$$x \cdot h' = (p \cdot g') \cdot h' = p \cdot g'h' = gh \cdot p = (ghg^{-1}) \cdot x.$$

Thus the right Schützenberger functor of $\mathcal{C}$ is in a sense superfluous.

This is most evident if S is commutative. Then $S/\mathcal{C}$ is commutative and all Schützenberger groups are abelian. Furthermore, $\theta_p : G^L(C) \longrightarrow G^R(C)$ depends only on the $\mathcal{C}$-class C of p: if $q = k \cdot p \in C$, then $g \cdot p = p \cdot g'$ implies

$$g \cdot q = gk \cdot p = kg \cdot p = k \cdot p \cdot g' = q \cdot g'$$

by (9.17), so that $\theta_p\, g \;=\; \theta_q\, g$. Hence θ_p may be denoted by θ_C. Moreover, the square

$$
\begin{array}{ccc}
G^L(C) & \xrightarrow{\;\theta_C\;} & G^R(C) \\
{\scriptstyle \gamma^L_{C,T}}\big\downarrow & & \big\downarrow{\scriptstyle \gamma^R_{T,C}} \\
G^L(CT) & \xrightarrow{\;\theta_{CT}\;} & G^R(CT) = G^L(TC)
\end{array}
\tag{9.18}
$$

commutes for all $T \in (S/\mathcal{C})^1$: if $p \in C$, $g \in G^L C$, and $t \in T$, then

$$
tp \cdot (\gamma^R_{T,C}\, \theta_C\, g) = t\,(p \cdot \theta_C\, g) = (g \cdot p)\, t
$$
$$
= \gamma^L_{C,T}\, g \cdot pt = tp \cdot (\theta_{CT}\, \gamma^L_{C,T}\, g)\,.
$$

Hence θ is a natural isomorphism of $\mathcal{G}^L$ onto $\mathcal{G}^R \circ \Theta$, where $\Theta \colon \mathcal{R}(S/\mathcal{C}) \longrightarrow \mathcal{L}(S/\mathcal{C})$ is the isomorphism $(C, T) \mapsto (T, C)$.

In general, given a congruence $\mathcal{C} \subseteq \mathcal{H}$ on S, then S can be reconstructed from $S/\mathcal{C}$ and the left Schützenberger functor of $\mathcal{C}$ alone, as a group coextension of $S/\mathcal{C}$. There is, however, a more elegant construction, essentially due to Leech [39], that treats both Schützenberger functors equally by combining them into a single two-sided 'functor'.

A *group valued difunctor* $\mathcal{G} = (G, \lambda, \rho)$ on a semigroup S (called a *compatible pair* in [39]) assigns a group G_a to each $a \in S$ and homomorphisms $\lambda_{t,a} : G_a \longrightarrow G_{ta}$ and $\rho_{a,t} : G_a \longrightarrow G_{at}$ to each $a \in S$ and $t \in S^1$, so that (G, λ) is a group valued functor on $\mathcal{R}(S)$ and (G, ρ) is a group valued functor on $\mathcal{L}(S)$. It is convenient to denote $\lambda_{t,a} g$ by ${}^t g$ and $\rho_{a,t} g$ by g^t.

Every group valued functor (G, γ) on the Leech category $\mathcal{D}(S)$ of S induces a difunctor (G, λ, ρ) on S, in which $\lambda_{t,a} = \gamma_{t,a,1}$, $\rho_{a,t} = \gamma_{1,a,t}$, and λ, ρ have the added property $({}^t g)^u = {}^t (g^u)$ for all g, t, u. Conversely, a group valued difunctor (G, λ, ρ) with this property expands to, and may be regarded as, a group valued functor (G, γ) on $\mathcal{D}(S)$, in which ${}^t g^u = ({}^t g)^u = {}^t (g^u)$ for all g, t, u.

Given a congruence $\mathcal{C} \subseteq \mathcal{H}$ on S, with Schützenberger functors $\mathcal{G}^L$ and $\mathcal{G}^R$, a group valued difunctor is constructed as follows. For each $C \in S/\mathcal{C}$, choose $p_C \in C$ and choose a group $G^d(C)$ and isomorphisms ζ_C and η_C:

$$
G^L(C) \xrightarrow{\;\zeta_C\;} G^d(C) \xrightarrow{\;\eta_C\;} G^R(C)
$$

such that $\eta_C \circ \zeta_C = \theta_{(p_C)}$. Let $G^d(C)$ act on C on the left and on the right by

$$
g \cdot x = \zeta_C^{-1} g \cdot x \quad \text{and} \quad x \cdot g = x \cdot \eta_C g
$$

for all $g \in G^d(C)$ and $x \in C$. Then $G^d(C)$ acts simply and transitively on C on the left and on the right. Moreover,

$$g \cdot p_C = \zeta_C^{-1} g \cdot p_C = p_C \cdot \theta_{p_C} \zeta_C^{-1} g = p_C \cdot \eta_C g = p_C \cdot g \qquad (9.19)$$

for all $g \in G^d(C)$, and (9.17) yields

$$g \cdot (x \cdot h) = \zeta_C^{-1} g \cdot (x \cdot \eta_C h) = (\zeta_C^{-1} g \cdot x) \cdot \eta_C h = (g \cdot x) \cdot h \qquad (9.20)$$

for all $g, h \in G^d(C)$ and $x \in C$.

Homomorphisms $\lambda_{T,C} \colon G^d(C) \longrightarrow G^d(TC)$ and $\rho_{C,U} \colon G^d(C) \longrightarrow G^d(CU)$ can then be defined for every $C \in S/\mathcal{C}$ and $T, U \in (S/\mathcal{C})^1$ so that the following squares commute:

$$
\begin{array}{ccc}
G^d(C) \xrightarrow{\;\eta_C\;} G^R(C) & \qquad & G^L(C) \xrightarrow{\;\zeta_C\;} G^d(C) \\
\lambda_{T,C} \downarrow \qquad \downarrow \gamma_{T,C}^R & \qquad & \gamma_{C,U}^L \downarrow \qquad \downarrow \rho_{C,U} \\
G^d(TC) \xrightarrow{\;\eta_{TC}\;} G^R(TC) & \qquad & G^L(CU) \xrightarrow{\;\zeta_{CU}\;} G^d(CU)
\end{array}
$$

for every $C \in S/\mathcal{C}$ and $T, U \in (S/\mathcal{C})^1$ (and ζ, η become natural isomorphisms). Then $\mathcal{G}^d(\mathcal{C}) = (G^d, \lambda, \rho)$ is a group valued difunctor on $S/\mathcal{C}$, the *Schützenberger difunctor* of $\mathcal{C}$. This difunctor depends on the choice of p, but different choices yield isomorphic difunctors. Moreover, $\mathcal{G}^d$ is *thin* in that $\lambda_{T,C}$ depends only on C and TC, and $\rho_{C,U}$ depends only on C and CU.

By definition,

$$^T g = \eta_{TC}^{-1} \gamma_{T,C}^R \eta_C g \;\text{ and }\; g^U = \zeta_{CU} \gamma_{C,U}^L \zeta_C^{-1} g$$

for all $g \in G^d(C)$. If $x \in C$, $g \in G^d(C)$, and $y \in D$, then

$$(g \cdot x) y = (\zeta_C^{-1} g \cdot x) y = \gamma_{C,D}^L \zeta_C^{-1} g \cdot xy \qquad (9.21)$$

$$= \zeta_{CD}^{-1} \rho_{C,D} g \cdot xy = \rho_{C,D} g \cdot xy = g^D \cdot xy;$$

if $x \in C$, $y \in D$, and $g \in G^d(D)$, then

$$x(y \cdot g) = x(y \cdot \eta_D g) = xy \cdot \gamma_{C,D}^R \eta_D g \qquad (9.22)$$

$$= xy \cdot \eta_{CD} \lambda_{C,D} g = xy \cdot \lambda_{C,D} g = xy \cdot {}^C g.$$

If S is commutative, then one may choose $G^d(C) = G^L(C)$, let ζ_C be the identity on $G^L(C)$, so that $\rho_{C,T} = \gamma_{C,T}^L$, and let $\eta_C = \theta_C$. Then $\lambda_{T,C} = \gamma_{C,T}^L$, by (9.18), and, in the difunctor $\mathcal{G}^d$, the subfunctors (G, λ) and (G, ρ) are both equal to $\mathcal{G}^L(\mathcal{C})$. Then S is a commutative group coextension of $S/\mathcal{C}$ by $\mathcal{G}^L(\mathcal{C})$, as defined in Chap. 1.

9.1.7 Leech Coextensions

Leech coextensions embody the main properties of the Schützenberger difunctors of congruences $\mathcal{C} \subseteq \mathcal{H}$, just as group coextensions embody the main properties of the Schützenberger functors of left coset congruences.

Let S be a semigroup and let $\mathcal{G} = (G, \lambda, \rho)$ be a group valued difunctor on S (perhaps, a group valued functor on $\mathcal{D}(S)$). We define a *Leech coextension* $E = (E, \pi, \cdot)$ of S by $\mathcal{G}$ as consisting of a semigroup E, a surjective homomorphism $\pi : E \longrightarrow S$ which partitions E into sets $E_a = \pi^{-1}a$ (one for each $a \in S$), and, for each $a \in S$, a simply transitive left action $\cdot$ of G_a on E_a and a simply transitive right action $\cdot$ of G_a on E_a, such that:

$$(g \cdot x)\, y = g^b \cdot xy \quad \text{and} \quad x\,(y \cdot h) = xy \cdot {}^a h \tag{9.23}$$

for all $g \in G_a$, $x \in E_a$, $y \in E_b$, $h \in G_b$, and $a, b \in S$;

$$(g \cdot x) \cdot h = g \cdot (x \cdot h) \tag{9.24}$$

for all $a \in S$, $g, h \in G_a$, and $x \in E_a$; and for each $a \in S$ there exists $p_a \in E_a$ such that

$$g \cdot p_a = p_a \cdot g \tag{9.25}$$

for all $g \in G_a$. In particular, E is a group coextension of S by (G, λ).

If $\mathcal{C}$ is a congruence on S contained in $\mathcal{H}$, then properties (9.21), (9.22), (9.20), and (9.19) show that S is a Leech coextension of $S/\mathcal{C}$ by the Schützenberger difunctor of $\mathcal{C}$.

Two Leech coextensions E and E' of S by a group valued difunctor $\mathcal{G}$ are *equivalent* if and only if there exists an isomorphism $\theta : E \longrightarrow E'$ (an *equivalence* of Leech coextension s) that preserves projection to S and the right and left actions of $\mathcal{G}$.

Leech coextensions are readily constructed up to equivalence by Schreier's method. Let $E = (E, \pi, \cdot)$ be a Leech coextension of a semigroup S by a group valued difunctor $\mathcal{G} = (G, \lambda, \rho)$ on S. Condition (9.25) provides $p_a \in E_a$ for each $a \in S$. As in every group coextension, every element x of E can be written uniquely in the form $x = g \cdot p_a$. In particular,

$$p_a\, p_b = s_{a,b} \cdot p_{ab},$$

where $s_{a,b} \in G_{ab}$ is the *factor set* of E (relative to the choice of p). Multiplication in E is now completely determined by s: for all $a, b \in S$, $g \in G_a$, and $h \in G_b$, (9.25), (9.23), and (9.24) yield

$$(g \cdot p_a)(h \cdot p_b) = g^b\, s_{a,b}\, {}^a h \cdot p_{ab}. \tag{9.26}$$

In the construction of E as a group coextension, comparing (9.26) to (9.14) and (9.15) shows that $\overset{h}{*}a$ depends only on a and h and that the commutation maps $h \mapsto \overset{h}{*}a$ are homomorphisms, when $\mathcal{C} \subseteq \mathcal{H}$.

Associativity for the multiplication in (9.26) is equivalent to

$$s_{a,bc} {}^a(h^c) {}^a s_{b,c} = s_{a,b}^c {}^{(a}h)^c s_{ab,c} \tag{9.27}$$

for all $a, b, c \in S$, $g \in G_a$, $h \in G_b$, and $k \in G_c$.

Conversely, if $\mathcal{G}$ is a group valued difunctor on S and s is a factor set that satisfies (9.27), then $E[s] = (E, \pi, \cdot)$, where

$$E = \{ (g, a) \mid a \in S, \ g \in g_a \}$$

with multiplication

$$(g, a)(h, b) = (g^b s_{a,b} {}^a h, \ ab),$$

together with projection $\pi : (g, a) \mapsto a$ to S and actions $g \cdot (h, a) = (gh, a)$, $(g, a) \cdot h = (gh, a)$ of G_a on E_a, is a Leech coextension of S by $\mathcal{G}$, in which $p_a = (1, a)$; and every Leech coextension of S by $\mathcal{G}$ with factor set s is equivalent to $E[s]$.

Moreover, two Leech coextensions E and E' of S by $\mathcal{G}$ with factor sets s and s' are equivalent if and only if there exists $u = (u_a)_{a \in S}$ such that $u_a \in G_a$ and

$$s'_{a,b} {}^a h \, u_{ab} = u_a^b s_{a,b} {}^a h \, {}^a u_b \tag{9.28}$$

for all $a, b \in S$ and $h \in G_b$.

Property (9.27) implies

$${}^a s_{b,c} s_{ab,c}^{-1} = s_{a,bc}^{-1} s_{a,b}^c \tag{9.29}$$

so that ${}^a(h^c)$ and $({}^a h)^c$ differ by an inner automorphism of G_{abc}. If they do not, then no Leech coextension of S by $\mathcal{G}$ can exist. If, however, $\mathcal{G}$ is a group valued functor on $\mathcal{D}(S)$, then ${}^a(h^c) = ({}^a h)^c$ whenever $h \in G_b$; the trivial factor set, $s_{a,b} = 1$ for all a, b, satisfies (9.27); and there exists at least one Leech coextension of S by $\mathcal{G}$, the *split* coextension.

If S is a monoid, then, as in [39], one may require Leech coextensions of S by $\mathcal{G}$ to be monoids as well. Choosing $p_1 = 1 \in E_1$ ensures a normalized factor set. Conversely, if S is a monoid and s is normalized, then $E[s]$ is a monoid, with $(1, 1)$ as identity element.

Let E be a Leech coextension of a semigroup S by a group valued difunctor $\mathcal{G} = (G, \lambda, \rho)$. By (9.26), E is commutative if and only if S is commutative and

$$g^b s_{a,b} {}^a h = h^a s_{b,a} {}^b g$$

whenever $g \in G_a$ and $h \in G_b$. This implies $s_{a,b} = s_{b,a}$ for all $a, b \in S$, and then $\lambda_{a,b}$ and $\rho_{b,a}$ differ by an inner automorphism of G_{ab}. If all groups G_a are abelian, then E is commutative if and only if S is commutative, $s_{a,b} = s_{b,a}$ for all $a, b \in S$, and $g^b = {}^b g$ for all $a, b \in S$ and $g \in G_a$; then E is a commutative group coextension of S by (G, ρ), as defined in Chap. 1.

9.1.8 Leech Cohomology

The Leech coextensions of S by $\mathcal{G}$ become simpler when all groups G_a are abelian. First, (9.24) and (9.25) imply

$$(g \cdot p_a) \cdot h = g \cdot (p_a \cdot h) = g \cdot (h \cdot p_a) = gh \cdot p_a = hg \cdot p_a = h \cdot (g \cdot p_a)$$

for all $g, h \in G_a$, so that the left and right actions of G_a on E_a coincide. Next, (9.27) and its consequence (9.29) imply

$$^a(h^c) = (^a h)^c \tag{9.30}$$

whenever $h \in G_b$, so that $\mathcal{G}$ may be regarded as a group valued functor (G, γ) on $\mathcal{D}(S)$, in which ${}^t g^u = ({}^t g)^u = {}^t (g^u)$ for all g, t, u. Conversely, (9.29) and (9.30) together imply (9.27).

Let $\mathcal{G}$ be an abelian group valued functor on $\mathcal{D}(S)$; write the groups G_a additively. If s is a factor set that satisfies (9.29):

$$^a s_{b,c} - s_{ab,c} + s_{a,bc} - s_{a,b}^c = 0 \tag{9.31}$$

for all $a, b, c \in S$, then $E[s] = (E, \pi, \cdot)$, where

$$E = \{ (g, a) \mid a \in S, \ g \in g_a \}$$

with multiplication

$$(g, a)(h, b) = (g^b + s_{a,b} + {}^a h, \ ab),$$

together with projection $\pi : (g, a) \mapsto a$ to S and actions $g \cdot (h, a) = (g + h, a) = (g, a) \cdot h$ of G_a on E_a, is a Leech coextension of S by $\mathcal{G}$; and every Leech coextension of S by $\mathcal{G}$ with factor set s is equivalent to $E[s]$. Moreover, two Leech coextensions E and E' of S by $\mathcal{G}$ with factor sets s and s' are equivalent if and only if there exists $u = (u_a)_{a \in S}$ such that $u_a \in G_a$ and

$$s_{a,b}' - s_{a,b} = u_a^b - u_{ab} + {}^a u_b \tag{9.32}$$

for all $a, b \in S$, by (9.28).

As suggested by (9.31) and (9.32), there is a cohomology that classifies Leech coextensions of S by an abelian group valued functor $\mathcal{G}$ on $\mathcal{D}(S)$. This is the *Leech cohomology* [40] of S with coefficients in $\mathcal{G}$, in which an n-cochain u assigns $u(a_1, a_2, \ldots, a_n) \in G_a$ to each $a_1, a_2, \ldots, a_n \in S$, where $a = a_1 a_2 \cdots a_n$, with coboundary

$$(\partial u)(a_1, a_2, \ldots, a_{n+1}) = {}^{a_1} u(a_2, \ldots, a_{n+1})$$
$$+ \sum_{1 \leq i \leq n} (-1)^i u(a_1, \ldots, a_i a_{i+1}, \ldots, a_{n+1})$$
$$+ (-1)^{n+1} u(a_1, a_2, \ldots, a_n)^{a_{n+1}}.$$

Thus (9.31) states that s is a 2-cocycle, and (9.32) states that two factor sets are equivalent if and only if they differ by a 2-coboundary. Hence the Leech cohomology *classifies* Leech coextensions of S by $\mathcal{G}$ [40].

If S is a monoid, as in [40], then one may require n-cochains to be *normalized*, in the sense that $u(a_1, a_2, \ldots, a_n) = 0$ if $a_i = 1$ for any i. Then $H^2(S, \mathcal{G})$ classifies the Leech coextensions of S by $\mathcal{G}$ that are also monoids. Moreover, Wells [54] has shown that the Leech cohomology of monoids is a particular case of Beck cohomology. This gives Leech cohomology a very strong claim to be the 'right' cohomology for monoids in general.

If S does not have an identity element, it seems likely that the Leech cohomology of S can be recovered from the normalized cohomology of S^1.

9.2 Monads and Algebras

This appendix recalls basic facts about monads and their algebras, following MacLane [42].

9.2.1 Adjunctions

An *adjunction* $(F, U, \eta, \varepsilon)$ from a category $\mathcal{X}$ to a category $\mathcal{C}$ consists of functors $F: \mathcal{X} \longrightarrow \mathcal{C}$, the *left adjoint* of U, and $U: \mathcal{C} \longrightarrow \mathcal{X}$, the *right adjoint* of F, and natural transformations $\eta: 1_{\mathcal{X}} \longrightarrow UF$, $\varepsilon: FU \longrightarrow 1_{\mathcal{C}}$ with the following universal properties:

(i) for every morphism $f: X \longrightarrow UC$ there is a unique morphism $\varphi: FX \longrightarrow C$ such that $f = U\varphi \circ \eta_X$;
(ii) for every morphism $\varphi: FX \longrightarrow C$ there is a unique morphism $f: X \longrightarrow UC$ such that $\varphi = \varepsilon_C \circ Ff$.

These universal properties are equivalent and provide, for every objects X and C of $\mathcal{X}$ and $\mathcal{C}$, a natural one-to-one correspondence between morphisms $f : X \longrightarrow UC$ and morphisms $\varphi : \mathbb{F}X \longrightarrow C$. In fact, the unique morphism $\varphi : FX \longrightarrow C$ such that $f = U\varphi \circ \eta_X$ is $\varepsilon_X \circ Ff$; and the unique morphism $f : X \longrightarrow UC$ such that $\varphi = \varepsilon_C \circ Ff$ is $U\varphi \circ \eta_X$.

In particular, η_X is the unique morphism $X \longrightarrow UFX$ such that

$$\varepsilon_{FX} \circ F\eta_X = 1_{FX} \tag{9.33}$$

and ε_C is the unique morphism $FUC \longrightarrow C$ such that

$$U\varepsilon_C \circ \eta_{UC} = 1_{UC}. \tag{9.34}$$

The 'free commutative semigroup' adjunction $(\mathbb{F}, \mathbb{U}, \eta, \varepsilon)$ from sets to commutative semigroups is a typical adjunction, in which $\mathbb{U}S$ is the underlying set of the commutative semigroup S (also denoted by S) and $\mathbb{F}X$ is the free commutative semigroup on the set X, which consists of all nonempty commutative words with letters in X. This adjunction is described in more detail in Sect. 2.3. The rest of this appendix recovers the category of commutative semigroups from parts of its adjunction.

9.2.2 Monads

A *monad* (T, η, μ) (formerly called a *triple*) in a category X is an abstract analogue of monoids that consists of a functor $T : \mathcal{X} \longrightarrow \mathcal{X}$, an 'identity' natural transformation $\eta : 1_{\mathcal{X}} \longrightarrow T$, and a natural transformation (a 'multiplication') $\mu : TT = T \circ T \longrightarrow T$, such that, for every object X of $\mathcal{X}$,

$$\mu_X \circ \eta_{TX} = 1_X = \mu_X \circ T\eta_X \text{ and } \mu_X \circ T\mu_X = \mu_X \circ \mu_{TX} : \tag{9.35}$$

$$
\begin{array}{ccc}
TX \xrightarrow{\eta_{TX}} TTX \xleftarrow{T\eta_X} TX & & TTTX \xrightarrow{T\mu_X} TTX \\
{}_{1_{TX}} \searrow \quad \downarrow {\mu_X} \quad \swarrow {}_{1_{TX}} & & {}_{\mu_{TX}} \downarrow \qquad \downarrow {\mu_X} \\
TX & & TTX \xrightarrow{\mu_X} TX.
\end{array}
$$

Every adjunction gives rise to a monad: if $(F, U, \eta, \varepsilon)$ is an adjunction from $\mathcal{X}$ to $\mathcal{C}$, then (T, η, μ), where

$$T = UF \text{ and } \mu = U\varepsilon F,$$

is a monad on $\mathcal{X}$: (9.35) follows from (9.33), (9.34), and the naturality of ε.

Dually, a *comonad* (V, ε, ν) (formerly called a *cotriple*) on a category $\mathcal{C}$ consists of a functor $V : \mathcal{C} \longrightarrow \mathcal{C}$ and natural transformations $\varepsilon : V \longrightarrow 1_{\mathcal{C}}$ and $\nu : V \longrightarrow VV$, such that, for every object C of $\mathcal{C}$,

$$\varepsilon_{VC} \circ \nu_C = 1_C = V\varepsilon_C \circ \nu_C \text{ and } V\nu_C \circ \nu_C = \nu_{VC} \circ \nu_C . \tag{9.36}$$

Every adjunction gives rise to a comonad: if $(F, U, \eta, \varepsilon)$ is an adjunction from $\mathcal{X}$ to $\mathcal{C}$, then (V, ε, ν), where

$$V = FU \text{ and } \nu = F\eta U ,$$

is a comonad on $\mathcal{C}$.

9.2.3 Algebras

Given a monad $T = (T, \eta, \mu)$ on a category $\mathcal{X}$, a *T-algebra* (A, α) consists of an object A of $\mathcal{X}$ and a morphism $\alpha : TA \longrightarrow A$, such that

$$\alpha \circ \eta_A = 1_A \text{ and } \alpha \circ T\alpha = \alpha \circ \mu_A : \tag{9.37}$$

$$
\begin{array}{ccc}
A \xrightarrow{\;\eta_A\;} TA & \qquad & TTA \xrightarrow{\;T\alpha\;} TA \\
{\scriptstyle 1_A}\searrow \quad \downarrow{\scriptstyle \alpha} & & {\scriptstyle \mu_A}\downarrow \qquad \downarrow{\scriptstyle \alpha} \\
A & & TA \xrightarrow{\;\alpha\;} A.
\end{array}
$$

For example, (TX, μ_X) is a T-algebra for every object X of $\mathcal{X}$, by (9.35).

A *morphism* $f : (A, \alpha) \longrightarrow (B, \beta)$ of T-algebras is a morphism $f : A \longrightarrow B$ such that

$$f \circ \alpha = \beta \circ Tf : \tag{9.38}$$

$$
\begin{array}{ccc}
TA & \xrightarrow{\;Tf\;} & TB \\
{\scriptstyle \alpha}\downarrow & & \downarrow{\scriptstyle \beta} \\
A & \xrightarrow{\;f\;} & B.
\end{array}
$$

T-algebras and their morphisms now constitute a category $\mathcal{X}^T$. This category comes with functors $F^T : \mathcal{X} \longrightarrow \mathcal{X}^T$ and $U^T : \mathcal{X}^T \longrightarrow \mathcal{X}$: for every object X and morphism f of $\mathcal{C}$, $F^T X$ is the T-algebra (TX, μ_X) and $F^T f = Tf$; for every object (A, α) and morphism f of $\mathcal{X}^T$, $U^T(A, \alpha) = A$ and $U^T f = f$. The functors F^T and U^T are part of an adjunction from $\mathcal{X}$ to $\mathcal{X}^T$ that gives rise to the given monad T.

Conversely, if an adjunction $(F, U, \eta, \varepsilon)$ from $\mathcal{X}$ to $\mathcal{C}$ gives rise to a monad T, then there is a unique functor $K : \mathcal{C} \longrightarrow \mathcal{X}^T$ such that $K \circ F = F^T$ and $U^T \circ K = U$:

$$
\begin{array}{ccccc}
\mathcal{X} & \xrightarrow{\ F\ } & \mathcal{C} & \xrightarrow{\ U\ } & \mathcal{X} \\
\downarrow{\scriptstyle 1_\mathcal{X}} & & \downarrow{\scriptstyle K} & & \downarrow{\scriptstyle 1_\mathcal{X}} \\
\mathcal{X} & \xrightarrow{\ F^T\ } & \mathcal{X}^T & \xrightarrow{\ U^T\ } & \mathcal{X}
\end{array}
$$

for every object C of $\mathcal{C}$, $KC = (UC, U\varepsilon_C)$; for every morphism φ of $\mathcal{C}$, $K\varphi = U\varphi$ [42].

In the adjunction $(F, U, \eta, \varepsilon)$ from $\mathcal{X}$ to $\mathcal{C}$, the functor U is *monadic* (also called *tripleable*) if and only if K is an isomorphism; by a blatant abuse of language, the category $\mathcal{C}$ is also said to be *monadic* (or *tripleable*) over $\mathcal{X}$.

Beck [6] and by Barr and Beck [5] give criteria for monadicity, mentioned in [42], which are used to prove that every variety of general algebraic systems (class definable by a set of identities) is monadic over the category of sets. Thus the categories of groups, rings, commutative rings, and so forth, are all monadic over the category of sets. In particular, the category of commutative semigroups is monadic over the category of sets.

9.3 Simplicial Objects

This appendix recalls basic facts about simplicial objects. MacLane [42] has a more comprehensive exposition.

9.3.1 Simplicial Sets

n-simplexes began life in Euclidean spaces as n-dimensional convex subsets with $n + 1$ extremal points or *vertices*, which can be numbered $0, 1, \ldots, n$. For example, triangles are 2-simplexes; tetrahedrons are 3-simplexes.

A Euclidean n-simplex has $n + 1$ *faces*, which are themselves $(n - 1)$-simplexes, face number i lying opposite vertex number i. Conversely an $(n - 1)$-simplex can be viewed as a *degenerate* (flattened) n-simplex S in n ways, by duplicating one vertex and placing the duplicate inside S.

The sets X_{n-1} and X_n of Euclidean $(n-1)$- and n-simplexes are thus related by $n+1$ *face maps* $d_i\colon X_n \longrightarrow X_{n-1}$ $(i = 0, 1, \ldots, n)$ and n *degeneracy maps* $s_i\colon X_{n-1} \longrightarrow X_n$ $(i = 0, 1, \ldots, n-1)$. These faces have the following properties, known as the *simplicial identities*

$$
\begin{aligned}
d_i \circ d_j &= d_{j-1} \circ d_i && \text{if } i < j, \\
d_i \circ s_j &= s_{j-1} \circ d_i && \text{if } i < j, \\
d_i \circ s_j &= 1_{X_{n-1}} && \text{if } i = j \text{ or if } i = j+1, \\
d_i \circ s_j &= s_j \circ d_{i-1} && \text{if } i > j+1, \\
s_i \circ s_j &= s_{j+1} \circ s_i && \text{if } i \leq j.
\end{aligned}
$$

Euclidean simplexes are the inspiration for the more abstract simplicial sets. A *simplicial set* X consists of a infinite sequence $X_0, X_1, \ldots, X_n, \ldots$ of sets and of *face maps* $d_i\colon X_n \longrightarrow X_{n-1}$ $(i = 0, 1, \ldots, n)$ and *degeneracy maps* $s_i\colon X_{n-1} \longrightarrow X_n$ $(i = 0, 1, \ldots, n-1)$ that satisfy the simplicial identities.

More generally, a *simplicial object* X in any category $\mathcal{X}$, consists of an infinite sequence $X_0, X_1, \ldots, X_n, \ldots$ of objects of $\mathcal{X}$ connected by face and degeneracy morphisms $d_i\colon X_{n+1} \longrightarrow X_n$ $(i = 0, 1, \ldots, n+1)$ and $s_i\colon X_n \longrightarrow X_{n+1}$ $(i = 0, 1, \ldots, n)$ that satisfy the simplicial identities.

A simplicial object X can be *augmented* by the addition of an object X_{-1} and morphism $d_0\colon X_0 \longrightarrow X_{-1}$, the *augmentation*, often denoted by ε.

Dually, a *cosimplicial object* in a category $\mathcal{X}$ consists of an infinite sequence $X_0, X_1, \ldots, X_n, \ldots$ of objects of $\mathcal{X}$ and of face and degeneracy morphisms $d_i\colon X_n \longrightarrow X_{n+1}$ $(i = 0, 1, \ldots, n+1)$ and $s_i\colon X_{n+1} \longrightarrow X_n$ $(i = 0, 1, \ldots, n)$ that satisfy the dual or *opposite simplicial identities*:

$$
\begin{aligned}
d_j \circ d_j &= d_i \circ d_{j-1} && \text{if } i < j, \\
s_j \circ d_i &= d_i \circ s_{j-1} && \text{if } i < j, \\
s_j \circ d_i &= 1 && \text{if } i = j \text{ or if } i = j+1, \\
s_j \circ d_i &= d_{i-1} \circ s_j && \text{if } i > j+1, \\
s_j \circ s_i &= s_i \circ s_{j+1} && \text{if } i \leq j.
\end{aligned}
$$

An *augmented cosimplicial object* is a cosimplicial object with an additional object X_{-1} and morphism $d_0\colon X_{-1} \longrightarrow X_0$.

9.3.2 The Simplicial Category

There is a more elegant definition of simplicial objects, based on the following category. The traditional *simplicial category* is the category Δ (denoted by Δ^+ in

[42]) in which an object is a set

$$[n] = \{0, 1, \ldots, n\}$$

(with $n + 1 > 0$ elements) and a morphism $\varphi: [m] \longrightarrow [n]$ is a *monotone* (order preserving) mapping ($i \leq j$ implies $\varphi(i) \leq \varphi(j)$).

There is for each $i \leq n$ one monotone surjection $\sigma_i: [n + 1] \longrightarrow [n]$ such that $\sigma_i \, i = \sigma_i \, (i + 1)$; σ_i is the i-th *degeneracy map* of Δ. There is for each $i \leq n + 1$ one monotone injection $\delta_i: [n] \longrightarrow [n + 1]$ such that $i \notin \delta_i [n]$; δ_i is the i-th *face map* of Δ.

Every morphism of Δ is a composite of face and degeneracy maps. The latter satisfy the opposite simplicial identities:

$$
\begin{aligned}
\delta_j \circ \delta_i &= \delta_i \circ \delta_{j-1} & &\text{if } i < j, \\
\sigma_j \circ \delta_i &= \delta_i \circ \sigma_{j-1} & &\text{if } i < j, \\
\sigma_j \circ \delta_i &= 1 & &\text{if } i = j \text{ or if } i = j + 1, \\
\sigma_j \circ \delta_i &= \delta_{i-1} \circ \sigma_j & &\text{if } i > j + 1, \\
\sigma_j \circ \sigma_i &= \sigma_i \circ \sigma_{j+1} & &\text{if } i \leq j.
\end{aligned}
$$

It turns out that Δ, as a category, has a presentation as generated by its face and degeneracy maps subject to the opposite simplicial identities: if a category $\mathcal{X}$ contains objects X_n ($n \geq 0$) and morphisms $d_i: X_n \longrightarrow X_{n+1}$ ($i = 0, 1, \ldots n$) and $s_i: X_{n+1} \longrightarrow X_n$ ($i = 0, 1, \ldots, n-1$) that satisfy the opposite simplicial identities, then there is a unique functor X of Δ into $\mathcal{X}$ such that $X[n] = X_n$, $X\delta_i = d_i$, and $X\sigma_i = s_i$, for all n and i.

Cosimplicial objects of a category $\mathcal{X}$ may now be defined as functors of Δ into $\mathcal{X}$. Similarly, simplicial objects of $\mathcal{X}$ may be defined as contravariant functors of Δ into $\mathcal{X}$.

Augmented simplicial objects and cosimplicial objects have a similar definition, using the following *augmented simplicial category* Δ^a (denoted by Δ in [42]). An object of Δ^a is a finite ordinal number

$$[n] = \{0, 1, \ldots, n - 1\}$$

(now with n elements). In particular, [0] is now the empty set, and an initial object of Δ^a. In Δ^a, a morphism from $[m]$ to $[n]$ is, as in Δ, a monotone mapping of $[m]$ into $[n]$. Thus, Δ is isomorphic to the full subcategory of Δ^a with objects [1], [2], ... Except for $\delta_0: [0] \longrightarrow [1]$, the face and degeneracy maps of Δ^a are the face and degeneracy maps of Δ.

An *augmented simplicial object* of a category $\mathcal{X}$ can now be defined as a contravariant functor X of Δ^a into $\mathcal{X}$; an *augmented cosimplicial object* of $\mathcal{X}$ can be defined as a (covariant) functor of Δ^a into $\mathcal{X}$.

9.3.3 The Classifying Simplicial Set

Eilenberg and MacLane [16] assigned a simplicial set to every simplicial monoid. This construction is used twice in Chap. 4.

As in the above, a simplicial monoid M consists of monoids M_0, M_1, ..., M_n, ... and of face and degeneracy maps that are monoid homomorphisms, $d_i : M_n \longrightarrow M_{n-1}$ ($i = 0, 1, \ldots, n$) and $s_i : M_{n-1} \longrightarrow M_n$ ($i = 0, 1, \ldots, n - 1$), and satisfy the simplicial identities.

The *classifying simplicial set* $V = V(M)$ (denoted by WM in [16], by $\overline{W}M$ in [10]) is constructed as follows.

First, V_0 is the cartesian product with zero terms, whose sole element is the empty sequence ().

For every $n \geq 0$, V_{n+1} is the cartesian product

$$V_{n+1} = M_n \times M_{n-1} \times \cdots \times M_0,$$

whose elements are all sequences $(x_n, \ldots, x_0)$ such that $x_i \in M_i$ for all i. We regard the sets V_n as pairwise disjoint.

The face and degeneracy maps d_i' and s_i' of V are constructed from the face and degeneracy maps d_i and s_i of M, so that d_i' shrinks $(x_n, \ldots, x_0) \in V_{n+1}$ using multiplication and s_i' inserts 1 in the i-th position:

$$d_0'(x_0) = d_1'(x_0) = (),\ \ s_0' = (1), \tag{9.39}$$

$$d_0'(x_n, \ldots, x_0) = (x_{n-1}, \ldots, x_0) \qquad \text{if } n > 0,$$

$$d_{i+1}'(x_n, \ldots, x_0) = (d_i x_n, \ldots, d_1 x_{n-i+1},$$
$$(d_0 x_{n-i})(x_{n-i-1}), x_{n-i-2}, \ldots, x_0) \qquad \text{if } i < n,$$

$$d_{n+1}'(x_n, \ldots, x_0) = (d_n x_n, \ldots, d_1 x_1) \qquad \text{if } n > 0,$$

$$s_0'(x_n, \ldots, x_0) = (1, x_n, \ldots, x_0),$$

$$s_{i+1}'(x_n, \ldots, x_0) = (s_i x_n, \ldots, s_0 x_{n-i},$$
$$1, x_{n-i-1}, \ldots, x_0) \qquad \text{if } i < n,$$

$$s_{n+1}'(x_n, \ldots, x_0) = (s_n x_n, \ldots, s_0 x_0, 1),$$

for all $(x_n, \ldots, x_0) \in V_{n+1}$. The proof that these maps satisfy the simplicial identities will not be inflicted upon the reader.

9.3.4 Cohomology

If there is a comonad (V, ε, ν) in a category $\mathcal{C}$, then every object C of $\mathcal{C}$ 'expands' to an augmented simplicial object C^s defined as follows.

The objects of C^s are $C^s_{-1} = C$, $C^s_0 = VC$, and, for every $n \geq 1$, $C^s_n = V^{n+1}C$, where $V^{n+1} = V \circ V \circ \cdots \circ V$ ($n + 1$ times).

The face maps are the natural transformations

$$\delta_i = V^i \varepsilon (V^{n-i}C) : C^s_{n+1} \longrightarrow C^s_n \ (i = 0, 1, \ldots, n).$$

The augmentation is $\varepsilon_C : VC \longrightarrow C$. The degeneracy maps are

$$\sigma_i = V^i \nu (V^{n-1-i}C) : C^s_n \longrightarrow C^s_{n+1} \ (i = 0, 1, \ldots, n - 1).$$

The simplicial identities are readily verified. This construction also follows from universal properties of Δ^a [42].

Any contravariant abelian group valued functor A on $\mathcal{C}$ carries C^s to an augmented cosimplicial abelian group, with objects AC^s_n ($n \geq -1$), face maps $d_i = A\delta_i$, and degeneracy maps $s_i = A\sigma_i$. A *coboundary* homomorphism $\delta : AC^s_n \longrightarrow AC^s_{n+1}$ is then defined by

$$\delta = d_0 - d_1 + d_2 - \cdots + (-1)^n d_n. \tag{9.40}$$

The simplicial identities imply $\delta \delta = 0$. The homology groups of the resulting cochain complex are the cohomology groups of C with coefficient in A:

$$H^n(C, A) = \operatorname{Ker} \delta / \operatorname{Im} \delta.$$

Beck cohomology in Chap. 2 and Gabriel-Zisman cohomology in Chap. 4 arise in precisely this fashion. The author does not have a partial augmented simplicial object for the symmetric cohomology.

Similarly, any (covariant) abelian group valued functor A on $\mathcal{C}$ then carries C^s to an augmented simplicial abelian group, yielding homology groups $H_n(C, A)$ of C with coefficient in A.

9.4 Monoidal Categories

This appendix contains basic definition and properties of monoidal categories and braided monoidal categories, following MacLane [42] and/or Joyal and Street [36].

9.4.1 Strict Monoidal Categories

A *multiplication* on a category C is a bifunctor $\cdot : C \times C \longrightarrow C$ and sends a pair (X, Y) of objects of C onto an object $XY = X \cdot Y$ and a pair (α, β) of morphisms of C onto a morphism $\alpha \cdot \beta$, so that

$$1_X \cdot 1_Y = 1_X \cdot_Y \quad \text{and} \quad (\alpha \cdot \beta) \circ (\gamma \cdot \delta) = (\alpha \circ \gamma) \cdot (\beta \circ \delta) \tag{9.41}$$

whenever defined. Common notations for $\cdot$ include $\square$ [42], $\otimes$ [10, 11, 36], and various other symbols.

A *strict monoidal category* $\mathfrak{M} = (\mathcal{M}, \cdot, I)$ consists of a category $\mathcal{M}$, a multiplication $\cdot$ on $\mathcal{M}$, and an *identity object* I of $\mathcal{M}$, such that

$$(XY)Z = X(YZ), \quad (\varphi \cdot \chi) \cdot \psi = \phi \cdot (\chi \cdot \psi), \quad \text{and} \tag{9.42}$$

$$XI = X = IX, \tag{9.43}$$

for all objects X, Y, Z and morphisms φ, χ, ψ of $\mathcal{M}$; so that $\cdot$ is associative and I is an identity element for objects of $\mathfrak{M}$.

9.4.2 General Monoidal Categories

Cartesian products and tensor products, however, are multiplications that are only associative up to natural isomorphisms. This suggests a more general concept.

A *monoidal category* $\mathfrak{M} = (\mathcal{M}, \cdot, I, \mathfrak{a}, \mathfrak{l}, \mathfrak{r})$ consists of a category $\mathcal{M}$, a multiplication $\cdot$ on $\mathcal{M}$, an identity object I of $\mathcal{M}$, and associativity, left identity, and right identity natural isomorphism s

$$\mathfrak{a} = \mathfrak{a}(X, Y, Z): (XY)Z \longrightarrow X(YZ),$$

$$\mathfrak{l} = \mathfrak{l}(X): IX \longrightarrow X, \quad \text{and}$$

$$\mathfrak{r} = \mathfrak{r}(X): XI \longrightarrow X,$$

with the following three compatibility properties:

$$\mathfrak{a}(X, Y, ZT) \circ \mathfrak{a}(XY, Z, T) \tag{9.44}$$

$$= 1_X \cdot \mathfrak{a}(Y, Z, T) \circ \mathfrak{a}(X, YZ, T) \circ \mathfrak{a}(X, Y, Z) \cdot 1_T :$$

$$((XY)Z)T \xrightarrow{\ \mathfrak{a}\ } (XY)(ZT) \xrightarrow{\ \mathfrak{a}\ } X(Y(ZT))$$

$$\mathfrak{a} \cdot 1 \searrow \qquad\qquad\qquad \nearrow 1 \cdot \mathfrak{a}$$

$$(X(YZ))T \xrightarrow{\ \mathfrak{a}\ } X((YZ)T)$$

$$1_X \cdot \mathfrak{l}(Y) \circ \mathfrak{a}(X, I, Y) = \mathfrak{r}(X) \cdot 1_Y : \tag{9.45}$$

$$1_X \cdot \mathfrak{l}(Y) \circ \mathfrak{a}(X, I, Y) = \mathfrak{r}(X) \cdot 1_Y :$$

$$
\begin{array}{ccc}
(XI)Y & \xrightarrow{\mathfrak{a}} & X(IY) \\
{\scriptstyle \mathfrak{r} \cdot 1} \searrow & & \swarrow {\scriptstyle 1 \cdot \mathfrak{l}} \\
 & XY &
\end{array}
$$

$$\mathfrak{l}(I) = \mathfrak{r}(I): I\,I \longrightarrow I, \tag{9.46}$$

for all objects X, Y, Z, T of $\mathcal{M}$.

Thus, a monoidal category is strict if and only if its isomorphisms $\mathfrak{a}$, $\mathfrak{l}$, and $\mathfrak{r}$ are identity isomorphisms.

Properties (9.44), (9.45), and (9.46) imply [42]:

$$\mathfrak{l}(YZ) \circ \mathfrak{a}(I, Y, Z) = \mathfrak{l}(Y) \cdot 1_Z, \tag{9.47}$$

$$1_X \cdot \mathfrak{r}(Y) \circ \mathfrak{a}(X, Y, I) = \mathfrak{r}(XY) : \tag{9.48}$$

$$
\begin{array}{ccc}
(IY)Z \xrightarrow{\mathfrak{a}} I(YZ) & \qquad & (XY)I \xrightarrow{\mathfrak{a}} X(YI) \\
{\scriptstyle \mathfrak{l} \cdot 1} \searrow \quad \swarrow {\scriptstyle \mathfrak{l}} & & {\scriptstyle \mathfrak{r}} \searrow \quad \swarrow {\scriptstyle 1 \cdot \mathfrak{r}} \\
YZ & & YZ
\end{array}
$$

for all objects X, Y.

More generally, the *Coherence Theorem* for monoidal categories states that the isomorphisms $\mathfrak{a}$, $\mathfrak{l}$, and $\mathfrak{r}$ are *coherent* in the following sense [37]: 'every' diagram that contains only instances of $\mathfrak{a}$, $\mathfrak{a}^{-1}$, $\mathfrak{l}$, $\mathfrak{l}^{-1}$, $\mathfrak{r}$, $\mathfrak{r}^{-1}$, and their products, is commutative. MacLane [42] gives a precise and careful statement of this result.

This Coherence Theorem can also be stated as follows: every monoidal category is equivalent (as a category) to a strict monoidal category.

9.4.3 Monoidal Functors

A *monoidal functor* $\mathfrak{F} = (\mathcal{F}, \zeta, f): \mathfrak{M} \longrightarrow \mathfrak{M}'$ of a monoidal category $\mathfrak{M} = (\mathcal{M}, \cdot, I, \mathfrak{a}, \mathfrak{l}, \mathfrak{r})$ into a monoidal category $\mathfrak{M}' = (\mathcal{M}', \cdot, I', \mathfrak{a}', \mathfrak{l}', \mathfrak{r}')$ consists of a functor $\mathcal{F}: \mathcal{M} \longrightarrow \mathcal{M}'$, a natural transformation $\zeta: \mathcal{F} - \cdot \mathcal{F} - \longrightarrow \mathcal{F}(- \cdot -)$, and a morphism $f: I' \longrightarrow \mathcal{F}I$, with the following properties: for all objects X, Y, Z of $\mathcal{M}$,

$$\mathcal{F}\mathfrak{a}(X, Y, Z) \circ \zeta(XY, Z) \circ (\zeta(X, Y) \cdot 1_{\mathcal{F}Z}) \tag{9.49}$$

$$= \zeta(X, YZ) \circ (1_{\mathcal{F}X} \cdot \zeta(Y, Z)) \circ \mathfrak{a}'(\mathcal{F}X, \mathcal{F}Y, \mathcal{F}Z) :$$

$$(\mathcal{F}X \cdot \mathcal{F}Y) \cdot \mathcal{F}Z \xrightarrow[\substack{\zeta \cdot 1 \\ \mathfrak{a}'}]{} \begin{array}{c} \mathcal{F}(XY) \cdot \mathcal{F}Z \xrightarrow{\zeta} \mathcal{F}((XY)Z) \searrow^{\mathcal{F}\mathfrak{a}} \\ \mathcal{F}X \cdot (\mathcal{F}Y \cdot \mathcal{F}Z) \xrightarrow{1 \cdot \zeta} \mathcal{F}X \cdot \mathcal{F}(YZ) \nearrow_{\zeta} \end{array} \mathcal{F}((XY)Z),$$

$$\mathcal{F}\mathfrak{l}(X) \circ \zeta(I, X) \circ (f \cdot 1_{\mathcal{F}X}) = \mathfrak{l}'(\mathcal{F}X) \text{ and} \qquad (9.50)$$

$$\mathcal{F}\mathfrak{r}(X) \circ \zeta(X, I) \circ (1_{\mathcal{F}X} \cdot f) = \mathfrak{r}'(\mathcal{F}X) :$$

$$
\begin{array}{ccc}
I' \cdot \mathcal{F}X \xrightarrow{\mathfrak{l}'} \mathcal{F}X & \qquad & \mathcal{F}X \cdot I' \xrightarrow{\mathfrak{r}'} \mathcal{F}X \\
f \cdot 1 \downarrow \quad \uparrow \mathcal{F}\mathfrak{l} & \qquad & 1 \cdot f \downarrow \quad \uparrow \mathcal{F}\mathfrak{r} \\
\mathcal{F}I \cdot \mathcal{F}X \xrightarrow{\zeta} \mathcal{F}(IX) & \qquad & \mathcal{F}X \cdot \mathcal{F}I \xrightarrow{\zeta} \mathcal{F}(XI).
\end{array}
$$

Monoidal functors compose as follows: if

$$\mathfrak{F} = (\mathcal{F}, \zeta, f) : \mathfrak{M} \longrightarrow \mathfrak{M}' \text{ and } \mathfrak{F}' = (\mathcal{F}', \zeta', f') : \mathfrak{M}' \longrightarrow \mathfrak{M}''$$

are monoidal functors, then $\mathfrak{F}' \circ \mathfrak{F} = (\mathcal{F}'', \zeta'', f'')$, where:

$$\mathcal{F}'' = \mathcal{F}' \circ \mathcal{F}; \qquad (9.51)$$

$$\zeta''(X, Y) = \mathcal{F}'\zeta(\mathcal{F}X, \mathcal{F}Y) + \zeta'(\mathcal{F}X, \mathcal{F}Y); \text{ and}$$

$$f'' = \mathcal{F}'f + f'.$$

is a monoidal functor $\mathfrak{F}'' : \mathfrak{M} \longrightarrow \mathfrak{M}''$.

A monoidal functor $\mathfrak{F} = (\mathcal{F}, \zeta, f)$ is *strong* if and only if $\mathcal{F}$ is an isomorphism (of categories), ζ is a natural isomorphism, and f is an isomorphism. A monoidal functor $\mathfrak{F} = (\mathcal{F}, \zeta, f)$ is *strict* if and only if $\mathcal{F}$, ζ, and f are all identity isomorphisms.

The 'strong' version of the Coherence Theorem for monoidal categories states that every monoidal category $\mathfrak{M}$ is equivalent (as a category) to a strict monoidal category $\mathfrak{M}'$, via strong monoidal functors $\mathfrak{M} \longrightarrow \mathfrak{M}'$ and $\mathfrak{M}' \longrightarrow \mathfrak{M}$ [42].

Monoidal functors communicate by monoidal natural transformations. (They don't have email.) Let $\mathfrak{F} = (\mathcal{F}, \zeta, f)$ and $\mathfrak{F}' = (\mathcal{F}', \zeta', f')$ be coterminal monoidal functors. A *monoidal natural transformation* $\tau : \mathfrak{F} \longrightarrow \mathfrak{F}'$ is an (ordinary) natural transformation $\tau : \mathcal{F} \longrightarrow \mathcal{F}'$ such that

$$\tau(XY) \circ \zeta(X, Y) = \zeta'(X, Y) \circ (\tau(X) \cdot \tau(Y)) \text{ and} \qquad (9.52)$$

$$\tau(X) \circ f = f' :$$

$$\begin{array}{ccc}
\mathcal{F}X \cdot \mathcal{F}Y & \xrightarrow{\;\zeta\;} & \mathcal{F}(XY) \\
{\scriptstyle \tau \cdot \tau} \downarrow & & \downarrow {\scriptstyle \tau} \\
\mathcal{F}'X \cdot \mathcal{F}'Y & \xrightarrow{\;\zeta'\;} & \mathcal{F}'(XY)
\end{array}
\qquad
\begin{array}{ccc}
I' & \xrightarrow{\;f\;} & \mathcal{F}I \\
{\scriptstyle 1} \downarrow & & \downarrow {\scriptstyle \tau} \\
I' & \xrightarrow{\;f'\;} & \mathcal{F}'I,
\end{array}$$

for all objects X, Y of $\mathcal{M}$.

9.4.4 Braided Monoidal Categories

A braided monoidal category is a monoidal category in which products commute up to isomorphism. Our definition follows Joyal and Street [36]; MacLane [42] has slightly different conditions, which do not seem to match accepted practice and do not match Calvo-Cervera and Cegarra [10]. Formally, then, a *braided monoidal category* $\mathfrak{M} = (\mathcal{M}, \cdot, I, \mathfrak{a}, \mathfrak{l}, \mathfrak{r}, \mathfrak{c})$, called a *braided tensor category* by Joyal and Street [36] consists of a monoidal category $(\mathcal{M}, \cdot, I, \mathfrak{a}, \mathfrak{l}, \mathfrak{r})$ together with a natural isomorphism $\mathfrak{c}$:

$$\mathfrak{c}(X, Y): XY \longrightarrow YX,$$

the *braiding*, such that:

$$\mathfrak{a}(Y, Z, X) \circ \mathfrak{c}(X, YZ) \circ \mathfrak{a}(X, Y, Z) \tag{9.53}$$
$$= 1_Y \cdot \mathfrak{c}(X, Z) \circ \mathfrak{a}(Y, X, Z) \circ \mathfrak{c}(X, Y) \cdot 1_Z :$$

$$\begin{array}{ccccc}
(XY)Z & \xrightarrow{\;\mathfrak{a}\;} & X(YZ) & \xrightarrow{\;\mathfrak{c}\;} & (YZ)X \\
{\scriptstyle \mathfrak{c} \cdot 1} \downarrow & & & & \downarrow {\scriptstyle \mathfrak{a}} \\
(YX)Z & \xrightarrow{\;\mathfrak{a}\;} & Y(XZ) & \xrightarrow{\;1 \cdot \mathfrak{c}\;} & Y(ZX), \text{ and}
\end{array}$$

$$\mathfrak{a}(Z, X, Y)^{-1} \circ \mathfrak{c}(XY, Z) \circ \mathfrak{a}(X, Y, Z)^{-1} \tag{9.54}$$
$$= \mathfrak{c}(X, Z) \cdot 1_Y \circ \mathfrak{a}(X, Z, Y)^{-1} \circ 1_X \cdot \mathfrak{c}(Y, Z) :$$

$$\begin{array}{ccccc}
X(YZ) & \xrightarrow{\;\mathfrak{a}^{-1}\;} & (XY)Z & \xrightarrow{\;\mathfrak{c}\;} & Z(XY) \\
{\scriptstyle 1 \cdot \mathfrak{c}} \downarrow & & & & \downarrow {\scriptstyle \mathfrak{a}^{-1}} \\
X(ZY) & \xrightarrow{\;\mathfrak{a}^{-1}\;} & (XZ)Y & \xrightarrow{\;\mathfrak{c} \cdot 1\;} & (ZX)Y.
\end{array}$$

These conditions imply

$$\mathfrak{l}(X) \circ \mathfrak{c}(X, I) = \mathfrak{r}(X) : \qquad\qquad (9.55)$$

$$
\begin{array}{ccc}
XI & \xrightarrow{\ \mathfrak{c}\ } & IX \\[4pt]
 & {}_{\mathfrak{r}}\searrow \quad \swarrow_{\mathfrak{l}} & \\[2pt]
 & X &
\end{array}
$$

A *braided monoidal functor* $\mathfrak{F} = (\mathcal{F}, \zeta, f)\colon \mathfrak{M} \longrightarrow \mathfrak{M}'$ of a braided monoidal category $\mathfrak{M} = (\mathcal{M}, \cdot, I, \mathfrak{a}, \mathfrak{l}, \mathfrak{r}, \mathfrak{c})$ into a braided monoidal category $\mathfrak{M}' = (\mathcal{M}', \cdot, I', \mathfrak{a}', \mathfrak{l}', \mathfrak{r}', \mathfrak{c}')$ is a monoidal functor $(\mathcal{F}, \zeta, f)$ (with properties (9.49) and (9.50)) that preserves braiding:

$$\mathcal{F}\mathfrak{c}(X, Y) \circ \zeta(X, Y) = \zeta(Y, X) \circ \mathfrak{c}'(\mathcal{F}X, \mathcal{F}Y) \qquad\qquad (9.56)$$

$$
\begin{array}{ccc}
\mathcal{F}X \cdot \mathcal{F}Y & \xrightarrow{\ \mathfrak{c}'\ } & \mathcal{F}Y \cdot \mathcal{F}X \\[4pt]
{}_{\zeta}\downarrow & & \downarrow_{\zeta} \\[4pt]
\mathcal{F}(XY) & \xrightarrow{\ \mathcal{F}\mathfrak{c}\ } & \mathcal{F}(YX).
\end{array}
$$

A *symmetric monoidal category* is a braided monoidal category $\mathfrak{M} = (\mathcal{M}, \cdot, I, \mathfrak{a}, \mathfrak{l}, \mathfrak{r}, \mathfrak{c})$ such that $\mathfrak{c}(X, Y)$ is an isomorphism and

$$\mathfrak{c}(Y, X) = \mathfrak{c}(X, Y)^{-1}, \qquad\qquad (9.57)$$

for every objects X, Y of $\mathcal{M}$. Property (9.49) then follows from (9.48).

The *Coherence Theorem* for monoidal categories extends to symmetric monoidal categories: the isomorphisms $\mathfrak{a}$, $\mathfrak{l}$, $\mathfrak{r}$, and $\mathfrak{c}$ of a symmetric monoidal category are coherent, so that 'every' diagram that contains only instances of $\mathfrak{a}$, $\mathfrak{a}^{-1}$, $\mathfrak{l}$, $\mathfrak{l}^{-1}$, $\mathfrak{r}$, $\mathfrak{r}^{-1}$, $\mathfrak{c}$, $\mathfrak{c}^{-1}$, and their products, is commutative.

A symmetric monoidal category $\mathfrak{M} = (\mathcal{M}, \cdot, I, \mathfrak{a}, \mathfrak{l}, \mathfrak{r}, \mathfrak{c})$ is *strictly symmetric* if and only if all its isomorphisms $\mathfrak{a}$, $\mathfrak{l}$, $\mathfrak{r}$, and $\mathfrak{c}$ are identity isomorphisms. Strictly symmetric monoidal categories are also called *commutative*, since $\mathfrak{c}(X, Y) = 1_{X \cdot Y}$ implies $Y \cdot X = X \cdot Y$ and, by naturality, $f \cdot g = g \cdot f$, for every objects X, Y and morphisms f, g.

The Coherence Theorem for symmetric monoidal categories can also be stated as follows: every symmetric monoidal category is equivalent (as a category) to a strict symmetric monoidal category.

The Coherence Theorem cannot apply in the same terms to braided monoidal categories in general; for example, diagrams

need not commute. Instead there is a test that decides when two composites of instances of $\mathfrak{a}, \mathfrak{a}^{-1}, \mathfrak{l}, \mathfrak{l}^{-1}, \mathfrak{r}, \mathfrak{r}^{-1}, \mathfrak{c}, \mathfrak{c}^{-1}$, and their products, are equal [42]. Although formulated for MacLane's slightly different definition, this test seems likely to apply to braided monoidal categories as defined here and in [36].

9.5   Modules

This appendix relates modules over S, or over its semigroup ring $\mathbb{Z}[S]$, to abelian group valued functors on S.

9.5.1   S-Modules

Given a commutative semigroup S, an *S-module* is an abelian group M together with a semigroup action of S on M (such that

$$a \cdot (x + y) = a \cdot x + a \cdot y \text{ and } a \cdot (b \cdot x) = ab \cdot x \qquad (9.58)$$

for all $a, b \in S$ and $x, y \in M$). If S is a monoid, then the action of S on M is also required to be a monoid action: $1 \cdot x = x$ for all $x \in M$, in addition to (9.58). A *homomorphism* of S-modules is an additive homomorphism φ that preserves the action of S: $\varphi (a \cdot x) = a \cdot \varphi (x)$ for all a, x.

If S does not have an identity element, then every semigroup action of S on M extends uniquely to a monoid action of S^1 on M. Hence the categories of S-modules and S^1-modules are isomorphic.

S-modules are related to modules over the semigroup ring $\mathbb{Z}[S]$ of S. Recall that $\mathbb{Z}[S]$ is the free abelian group on the set S, with multiplication induced by S: an element of $\mathbb{Z}[S]$ is a linear combination $\sum_{a \in S} n_a a$ with integer coefficients $n_a \in \mathbb{Z}$ such that $n_a \neq 0$ for only finitely many $a \in S$; the elements of $\mathbb{Z}[S]$ multiply by:

$$\Big(\sum_{a \in S} n_a a\Big)\Big(\sum_{b \in S} m_b b\Big) = \sum_{a,b \in S} n_a m_b ab = \sum_{c \in S}\Big(\sum_{ab=c} n_a m_b c\Big).$$

If S has an identity element, then so does $\mathbb{Z}[S]$.

Every semigroup action of S on an abelian group M extends uniquely to a ring action of $\mathbb{Z}[S]$ on M, making M a $\mathbb{Z}[S]$-module. Hence the categories of S-modules and $\mathbb{Z}[S]$-modules are isomorphic (and isomorphic to the categories of S^1-modules and $\mathbb{Z}[S^1]$-module s).

9.5.2 Quasiconstant Functors

An abelian group valued functor $\mathcal{G} = (G, \gamma)$ on S is *quasiconstant* if and only if (i) $G_a = G_b$ for all $a, b \in S$ and (ii) $\gamma_{a,t} = \gamma_{b,t}$ for all $a, b \in S$ and $t \in S^1$; then $\mathcal{G}$ is *quasiconstant at* $G = G_a = G_b$.

Quasiconstant functorss do not lend themselves to universal coefficients theorems, and were not included in Sect. 7.4.

A *homomorphism* of quasiconstant functors from $\mathcal{G} = (G, \gamma)$ to $\mathcal{G}' = (G', \gamma')$, where $\mathcal{G}$ is quasiconstant at G and $\mathcal{G}'$ is quasiconstant at G', is a homomorphism $\varphi \colon G \longrightarrow G'$ such that $\varphi \circ \gamma_{a,t} = \gamma'_{a,t} \circ \varphi$ for all a and t (so that φ is a constant natural transformation, $\varphi(a) = \varphi$ for all $a \in S$).

The main result in this section is:

The category of S-modules is isomorphic to the category of quasiconstant functors on S.

The isomorphisms are as follows. Given an S-module M, define $\mathbb{A}(M) = (G, \gamma)$ by: $G_a = M$ and $\gamma_{a,t} x = t \cdot x$ for all $a \in S$ and $x \in M$. Then $\mathbb{A}(M)$ is an abelian group valued functor on S and is quasiconstant at M.

Conversely, given an abelian group valued functor $\mathcal{G} = (G, \gamma)$ on S that is quasiconstant at some abelian group M, an action of S on M is well defined by: $t \cdot x = \gamma_{a,t} x$, for any $a \in S$, and makes M into an S-module $\mathbb{M}(\mathcal{G})$.

9.5.3 Conclusions

First, abelian group valued functors on S are much more general than S-modules or $\mathbb{Z}[S]$-modules.

Second, the André-Quillen cohomology [3, 46] of $\mathbb{Z}[S]$ is not an adequate substitute for the cohomology of S: since $\mathbb{Z}[S]$-modules provide its coefficients, it cannot classify all commutative semigroups, only those commutative semigroups whose Schützenberger functors are quasiconstant. (This may require further investigations.)

References

1. W.W. Adams, P. Loustaunau, *An Introduction to Gröbner Bases* (Amer. Math. Soc., Providence, 1994)
2. M. André, *Méthode simpliciale en algèbre homologique et algèbre commutative*. Lecture Notes No.32 (Springer, Berlin, 1967)
3. M. André, *Homologie des algèbres commutatives* (Springer, Berlin, 1974)
4. D.D. Arendt, C.J. Stuth, On the structure of commutative periodic semigroups. Pac. J. Math. **35**, 1–6 (1970)
5. M. Barr, J. Beck, *Homology and Standard Constructions*. Lecture Notes No.80 (Springer, Berlin, 1969), pp 235–335
6. J. Beck, Triples, algebras, and cohomology. Doct. Diss., Columbia Univ. (1967)
7. T. Becker, V. Weispfenning, *Gröbner Bases* (Springer, Berlin, 1998)
8. H. Brandt, Uber die Axiome des Gruppoids. Vierteljscher. Naturforsch. Ges. Zurich **85**, 95–104 (1940)
9. K.S. Brown, *Cohomology of Groups* (Springer, Berlin, 1982)
10. M. Calvo-Cervera, A.M. Cegarra, A cohomology theory for commutative monoids. Mathematics **3**, 1001–1031 (2015)
11. M. Calvo-Cervera, A.M. Cegarra, B.A. Heredia, On the third cohomology group of commutative monoids. Semigroup Forum (2016). https://doi.org/10.1007/s00233-015-9696-2
12. A.M. Cegarra, F. Khadmaladze, Homotopy classification of graded Picard categories. Adv. Math. **213**, 644–686 (2007)
13. A.H. Clifford, G.B. Preston, *The Algebraic Theory of Semigroups*, vol. I (Amer. Math. Soc., Providence, 1961)
14. L.E. Dickson, Finiteness of the odd perfect and primitive abundant numbers with n distinct prime factors. Am. J. Math. **35**, 413–422 (1913)
15. J. Duskin, *Simplicial Methods and the Interpretation of "Triple" Cohomology*, vol. 163 (Mem. Amer. Math. Soc., Providence, 1975)
16. S. Eilenberg, S. MacLane, On the groups $H(\Pi, n)$, I. Ann. Math. **58**, 55–106 (1953)
17. P. Gabriel, M. Zisman, *Calculus of Fractions and Homotopy Theory* (Springer, Berlin, 1967)
18. P.A Grillet, Left coset extensions. Semigroup Forum **7**, 200–263 (1974)
19. P.A. Grillet, A completion theorem for finitely generated commutative semigroups. J. Algebra **34**, 25–53 (1975)
20. P.A. Grillet, Primary semigroups. Michigan Math. J. **22**, 321–326 (1975)
21. P.A. Grillet, Commutative semigroup cohomology. Semigroup Forum **43**, 247–252 (1991)

© The Author(s), under exclusive license to Springer Nature Switzerland AG 2022
P. A. Grillet, *The Cohomology of Commutative Semigroups*, Lecture Notes
in Mathematics 2307, https://doi.org/10.1007/978-3-031-08212-2

22. P.A. Grillet, The commutative cohomology of nilsemigroups. J. Pure Appl. Algebra **82**, 233–251 (1992)
23. P.A. Grillet, Some congruences on free commutative semigroups. Acta Sci. Math. (Szeged) **58**, 3–23 (1993)
24. P.A. Grillet, A short proof of Rédei's Theorem. Semigroup Forum **46**, 126–127 (1993)
25. P.A. Grillet, Commutative semigroup cohomology. Comm. Algebra **23**(10), 3573–3587 (1995)
26. P.A. Grillet, The commutative cohomology of finite semigroups. J. Pure Appl. Algebra **102**, 25–47 (1995)
27. P.A. Grillet, Partially free commutative semigroups. Acta Sci. Math. (Szeged) **61**, 155–169 (1995)
28. P.A. Grillet, Cocycles in commutative semigroup cohomology. Commun. Algebra **25**(11), 3427–3462 (1997)
29. P.A. Grillet, Commutative semigroups with zero cohomology. Acta Sci. Math. (Szeged) **63**, 463–469 (1997)
30. P.A. Grillet, Commutative nilsemigroups with zero cohomology. Semigroup Forum **62**, 66–78 (2001)
31. P.A. Grillet, *Commutative Semigroups* (Springer, Berlin, 2001)
32. P.A. Grillet, Four-cocycles in commutative semigroup cohomology. Semigroup Forum (2020). https://doi.org/10.1007/s00233-019-10046-9
33. P.A. Grillet, The commutative cohomology of nilmonoids, revisited. Semigroup Forum (2020). https://doi.org/10.1007/s00233-020-10107-4
34. P.A. Grillet, Commutative monoid homology. Semigroup Forum (2021). https://doi.org/10.1007/s00233-021-10194-x
35. P.A. Grillet, The inheritance of symmetric conditions in commutative semigroup cohomology. Semigroup Forum **104**, 72–87 (2022). https://doi.org/10.1007/s00233-021-10239-1
36. S. Joyal, R. Street, Braided tensor categories. Adv. Math. **102**, 20–78 (1993) https://doi.org/10.1006/aima.1993.1055
37. G.M. Kelly, On Mac Lane's conditions for coherence of natural associativities, commutativities, etc. J. Algebra **1**, 397–402 (1964)
38. R. Kurdiani, T. Pirashvili, Functor homology and homology of commutative monoids. Semigroup Forum **92**, 102–120 (2016)
39. J. Leech, $\mathcal{H}$-coextensions of monoids, in *Mem. Amer. Math. Soc.*, vol. 157 (1975), pp 1–66
40. J. Leech, The cohomology of monoids. Unpublished lecture notes (1976)
41. S. MacLane, *Homology* (Springer, Berlin, 1963)
42. S. MacLane, *Categories for the working mathematician*, 2nd edn. (Springer, Berlin, 1998)
43. I.S. Ponizovsky, A remark on commutative semigroups. Dokl. Akad. Nauk SSSR **142**, 1258–1260 (1962)
44. G.B. Preston, Rédei's characterization of congruences on finitely generated free commutative semigroups. Acta Sci. Math. Acad. Sci. Hungar. **26**, 337–342 (1975)
45. D.G. Quillen, *Homotopical Algebra*. Lecture Notes No.43 (Springer, Berlin, 1967)
46. D.G. Quillen, On the (co)homology of commutative rings, in *Proc. Symposia Pure Math.* No.17. Applications of Categorical Algebra (Amer. Math. Soc., Providence, 1970)
47. L. Rédei, Die Verallgemeinerung der Schreierschen Erweiterungtheorie. Acta Sci. Math. (Szeged) **14**, 252–273 (1952)
48. L. Rédei, The theory of finitely generated commutative semigroups. Akad. Kiadó, Budapest (1956). English translation, Pergamon Press, Oxford (1965).
49. J.E. Roos, Sur les foncteurs dérivés de projlim. Applications. C. R. Acad. Sci. Paris **252**, 3702–3704 (1961).
50. J.C. Rosales, Function minimum associated with to a congruence on integral n-tuple space. Semigroup Forum **51**, 87–95 (1995)
51. J.C. Rosales, P.A. García-Sánchez, *Finitely Generated Commutative Semigroups* (Nova Science Publishers, Commack, 1999)

52. O. Schreier, Über die Erweiterungen von Gruppen. I. Monatsch. Math. Phys. **34**, 165–180 (1926).
53. C.A. Weibel, *An Introduction to Homological Algebra* (Cambridge Univ. Press, Cambridge, 1994)
54. C. Wells, Extension theories for monoids. Semigroup Forum **16**, 13–35 (1978)

Index

A

abelian group object, 10
  morphism of, 10
abelian groupoid, 60
abelian group valued functor, 3
  almost constant, 122
  constant, 3, 55, 122
  quasiconstant, 166
  selective, 105, 122
  semiconstant, 104, 122
  surjecting, 8
  thin, 3
action
  monoid, 165
  semigroup, 165
adjunction, 153
  of commutative semigroups, 18
admissible ordering, 88
almost constant functor, 123
André-Quillen cohomology, vii, viii, 166
augmented
  cosimplicial object, 157
  simplicial category, 156
  simplicial object, 155, 157
augmentation, 156, 159
antichain, 88

B

base of reduced abel. groupoid, 64, 81
basic
  condition, 102
  relation, 102
  relationship, 136
  symmetry property, 136

basis
  that consists of defining vectors, 105
  defining, 104
  defining vector, 102
  of free commutative semigroup, 87
  Gröbner, 89
  optimal, 105
  reaching, 105
  standard, 111
  of symmetric set, 110
Beck
  coboundary, 2
  cochain, 21
  cocycle, 24
  cohomology (*see* Beck cohomology)
  extension, 13
    of commutative semigroups, 17-19
Beck cohomology, vii, 9–25, 117
  chains in, 117
  of comm. semigroup, 18–25
    properties, 25
  of comonad, 159
  compared to symmetric coh., 32–38
  general definition, 12
  general properties, 13
boundary
  relative to B, 125
  symmetric, 117, 125
  of symmetric chain, 117
braided
  abelian $\otimes$-groupoid, 77
  $\otimes$ functor, 79
  monoidal abelian groupoid (*see* braided
    monoidal abelian groupoid)
  monoidal category, 60, 163

© The Author(s), under exclusive license to Springer Nature Switzerland AG 2022
P. A. Grillet, *The Cohomology of Commutative Semigroups*, Lecture Notes
in Mathematics 2307, https://doi.org/10.1007/978-3-031-08212-2

LECTURE NOTES IN MATHEMATICS

Editors in Chief: J.-M. Morel, B. Teissier;

Editorial Policy

1. Lecture Notes aim to report new developments in all areas of mathematics and their applications – quickly, informally and at a high level. Mathematical texts analysing new developments in modelling and numerical simulation are welcome.

   Manuscripts should be reasonably self-contained and rounded off. Thus they may, and often will, present not only results of the author but also related work by other people. They may be based on specialised lecture courses. Furthermore, the manuscripts should provide sufficient motivation, examples and applications. This clearly distinguishes Lecture Notes from journal articles or technical reports which normally are very concise. Articles intended for a journal but too long to be accepted by most journals, usually do not have this "lecture notes" character. For similar reasons it is unusual for doctoral theses to be accepted for the Lecture Notes series, though habilitation theses may be appropriate.

2. Besides monographs, multi-author manuscripts resulting from SUMMER SCHOOLS or similar INTENSIVE COURSES are welcome, provided their objective was held to present an active mathematical topic to an audience at the beginning or intermediate graduate level (a list of participants should be provided).

   The resulting manuscript should not be just a collection of course notes, but should require advance planning and coordination among the main lecturers. The subject matter should dictate the structure of the book. This structure should be motivated and explained in a scientific introduction, and the notation, references, index and formulation of results should be, if possible, unified by the editors. Each contribution should have an abstract and an introduction referring to the other contributions. In other words, more preparatory work must go into a multi-authored volume than simply assembling a disparate collection of papers, communicated at the event.

3. Manuscripts should be submitted either online at www.editorialmanager.com/lnm to Springer's mathematics editorial in Heidelberg, or electronically to one of the series editors. Authors should be aware that incomplete or insufficiently close-to-final manuscripts almost always result in longer refereeing times and nevertheless unclear referees' recommendations, making further refereeing of a final draft necessary. The strict minimum amount of material that will be considered should include a detailed outline describing the planned contents of each chapter, a bibliography and several sample chapters. Parallel submission of a manuscript to another publisher while under consideration for LNM is not acceptable and can lead to rejection.

4. In general, **monographs** will be sent out to at least 2 external referees for evaluation.

   A final decision to publish can be made only on the basis of the complete manuscript, however a refereeing process leading to a preliminary decision can be based on a pre-final or incomplete manuscript.

   Volume Editors of **multi-author works** are expected to arrange for the refereeing, to the usual scientific standards, of the individual contributions. If the resulting reports can be forwarded to the LNM Editorial Board, this is very helpful. If no reports are forwarded or if other questions remain unclear in respect of homogeneity etc, the series editors may wish to consult external referees for an overall evaluation of the volume.

5. Manuscripts should in general be submitted in English. Final manuscripts should contain at least 100 pages of mathematical text and should always include

 – a table of contents;
 – an informative introduction, with adequate motivation and perhaps some historical remarks: it should be accessible to a reader not intimately familiar with the topic treated;
 – a subject index: as a rule this is genuinely helpful for the reader.
 – For evaluation purposes, manuscripts should be submitted as pdf files.

6. Careful preparation of the manuscripts will help keep production time short besides ensuring satisfactory appearance of the finished book in print and online. After acceptance of the manuscript authors will be asked to prepare the final LaTeX source files (see LaTeX templates online: https://www.springer.com/gb/authors-editors/book-authors-editors/manuscriptpreparation/5636) plus the corresponding pdf- or zipped ps-file. The LaTeX source files are essential for producing the full-text online version of the book, see http://link.springer.com/bookseries/304 for the existing online volumes of LNM). The technical production of a Lecture Notes volume takes approximately 12 weeks. Additional instructions, if necessary, are available on request from lnm@springer.com.

7. Authors receive a total of 30 free copies of their volume and free access to their book on SpringerLink, but no royalties. They are entitled to a discount of 33.3 % on the price of Springer books purchased for their personal use, if ordering directly from Springer.

8. Commitment to publish is made by a *Publishing Agreement*; contributing authors of multiauthor books are requested to sign a *Consent to Publish form*. Springer-Verlag registers the copyright for each volume. Authors are free to reuse material contained in their LNM volumes in later publications: a brief written (or e-mail) request for formal permission is sufficient.

Addresses:
Professor Jean-Michel Morel, CMLA, École Normale Supérieure de Cachan, France
E-mail: moreljeanmichel@gmail.com

Professor Bernard Teissier, Equipe Géométrie et Dynamique,
Institut de Mathématiques de Jussieu – Paris Rive Gauche, Paris, France
E-mail: bernard.teissier@imj-prg.fr

Springer: Ute McCrory, Mathematics, Heidelberg, Germany,
E-mail: lnm@springer.com

Printed in the United States
by Baker & Taylor Publisher Services